INTRODUCTION
TO
Optical Mineralogy and Petrography

The Practical Methods of
Identifying Minerals
in Thin Section
With the
Microscope

and

The Principles Involved in
The Classification of Rocks

By

M. G. EDWARDS, A. M.
Instructor in Geology and Mineralogy
Case School of Applied Science.

CLEVELAND, OHIO,
1916.

The Gardner Printing Co.
Cleveland
1916

PREFACE.

IN THE preparation of this volume the writer has attempted to gather together and systematize in a manner accessible for ready reference those facts which are essential to a field geologist or to a mining engineer in an understanding of the fundamental principles involved in the classification and identification of rocks. In the field, a preliminary classification is usually made by macroscopic means. However, it is often necessary to make a more careful classification by a microscopic examination of a thin section of the minerals comprising the rock mass. To do this successfully requires a knowledge of the application of light to crystalline substances.

This volume differs from most of the reference and text books relating to this subject in that it incorporates in one volume the elements of optical mineralogy and the elements of petrography. In Part One, eight general operations for the determination of unknown minerals in thin section are described, prefaced by a short summary of the principles of optics which apply to the transmission of polarized light through minerals. Descriptions of fifty-eight of the most common of the rock-making minerals are given, special attention being given to the criteria for the determination of these minerals in thin section. Their form, cleavage, twinning, color, refringence, birefringence, extinction angles, pleochroism, absorption, optical character, inclusions, alterations, occurrences, uses, and differentiation from similar minerals, are all discussed whenever applicable. An elementary knowledge of crystallography and descriptive mineralogy is assumed.

In Part Two, the principles of petrography are discussed briefly. Attention is given to the classification and description of the more important igneous rock types.

Following Iddings, Winchell, and other American petrographers, the symbols X, Y, and Z, are here employed in referring to the axes of ether elasticity, instead of the German a, b, and c, used in many text and reference books. This is done to avoid confusion, especially in conversation or discussion, with the crystallographic axes.

The writer is indebted to Professor Frank R. Van Horn for suggestions. Among the reference and text books most frequently consulted the writer wishes to acknowledge Winchell's "Elements of Optical Mineralogy," Johannsen's "Manual of Petrographic Methods," Luquer's "Minerals in Rock Sections," Rogers's "Study of Minerals," Findlay's "Igneous Rocks," Kemp's "Handbook of Rocks," Ries' and Watson's "Engineering Geology," and Farrell's "Practical Field Geology."

M. G. EDWARDS.

Cleveland, Ohio, February, 1916.

TABLE OF CONTENTS

The Nature of Light — Isotropic and Aniso-
tropic Media — Uniaxial and Biaxial Crystals
— Index of Refraction — Double Refraction —
Interference — Polarization.

Microscope — Nicol Prisms — Condensing Lens
— Cross Hairs — Stage — Mirror — Objective
— Bertrand Lens — Ocular Micrometer — Ad-
justment Screws.

1. By the General Physical Properties; 2. By
the Relative Refractive Index — Method of Duc
de Chaulnes — Immersion M e t h o d — Becke
Method — Scale of Refringence; 3. By the Birefringence — Interference C o l o r s — Axes of
Ether Vibration — Optic Plane — Scale of Birefringence.

4. By Axial Interference Figures — Uniaxial

INTRODUCTION.

THE TERM Petrology is derived from the two Greek words *petros* (rock) and *logos* (discourse), from which the modern definition, the science or treatise of rocks, has been evolved. The term has a wide scope, and embraces not only the study of the origin and transformation of rocks but a consideration of their mineral composition, classification, description and identification based upon either megascopic or microscopic characteristics.

Petrology may be subdivided into the following special studies:

Petrogeny, which is concerned with the origin of rocks, and

Petrography, which is concerned with the systematic classification and description of rocks megascopically and microscopically. It is the latter phase of the subject which is dealt with chiefly in the following notes.

Petrography may be divided for the sake of convenience into megascopic petrography and microscopic petrography, depending upon whether or not the student is basing his identification, classification and description upon a study of the rock in hand specimen or in thin section with the aid of the polarizing microscope.

The use of the polarizing microscope necessarily entails a brief review of the elements of optics and a consideration of the application of polarizing light to crystalline substances. This is a special study in itself, and is called Optical Mineralogy. Assuming that the student has had little or no previous experience with the study of the optical properties of minerals, a short review of the optical characters of the more important rock-making minerals is given. Special attention is given to the criteria for the determination of the mineral in thin section and diagnostics for the differentiation of the mineral from similar minerals.

History of Petrography.—Great advances in the knowledge of mineralogy marked the latter half of the eighteenth century. Incidentally there followed several attempts to classify rocks, which resulted in 1787 in the publication of two classifications by Karl Haidinger (Vienna) and A. G. Werner (Dresden). Werner's classification was stratigraphic rather than petrographic, but he described rocks in terms of mineralogical composition and physical characteristics, and he differentiated between essential and accessory minerals.

In 1801, Abbé R. J. Hauy (Paris), a mineralogist, published the first systematic classification, and his "Traité de mineralogie" with subsequent revisions remained a classic for a long period. He distinguished five classes of rocks: stony and saline, combustible nonmetallic, metallic, rocks of an igneous or aqueous origin, and volcanic rocks.

John Pinkerton (England) in 1811 published a "Petrology, a Treatise on Rocks," of 1200 pages. In view of the fact that natural history was divided into three kingdoms—the animal, vegetable, and mineral—he believed it the most natural classification to subdivide the mineral kingdom into provinces and domains. Accordingly he introduced the following three provinces: Petrology, the knowledge of rocks or stones in large masses; Lithology, the knowledge of gems and small stones, and Metallogy, the knowledge of metals. Pinkerton's volume was lightly regarded even by his contemporaries.

Cordier (France) in 1815 classified rocks as feldspathic or pyroxenic, and made subdivisions according to texture.

Karl von Leonhard (Heidelberg) in 1823 and Alexandre Brongniart in 1827 proposed systems which mark the real origin of systematic petrography. Mineral composition was the chief factor in the classification. The former established four divisions: heterogeneous rocks, homogeneous rocks, fragmental rocks, and loose rocks. The latter made only two classes: homogeneous rocks and heterogeneous rocks.

Hermann Abich in 1841 made a classification of the eruptive rocks according to the content of the various feldspars.

The term petrography was perhaps first used by Carl Friedrich Naumann, who in 1850 published his "Lehrbuch der Geognosie," in which he divided all rock classes into crystalline rocks, clastic rocks, and rocks which are neither crystalline nor clastic. In a later revision he recognized only two classes: the original, and the derived.

Several classifications were presented in the next few decades by von Cotta (1855), Senft (1857), Blum (1860), Roth (1861), Scheerer (1864), Ferdinand Zirkel (1866), and F. von Richthofen (1868), based upon mineral constitution, chemical composition, structure, and texture, with an increasing tendency to emphasize the importance of mineral composition.

A new era in the development of petrography dawned with the introduction of the polarizing microscope. With the greater knowledge of mineral composition and texture thus revealed, the old schemes were discarded or radically revised, new terms introduced, and the nomenclature became rapidly more complex. Although Henry Clifton Sorbey (England) perhaps first used the microscope in the determination of rock sections, it was not until the decade between 1870 and 1880 that microscopic methods began to exert a controlling influence in the development

of the science. Zirkel in 1873 produced "Die mikroskop-
ische Beschaffenheit der Mineralien und Gesteine," which
shows a remarkable and significant advance in the progress of petrography in the eight years following the publication of his "Lehrbuch." He dealt chiefly with feldspathic, massive, composite, and nonclastic rocks.

In France in 1879 the "Mineralogie micrographique,"
by F. Fouque and A. Michel Levy, appeared. Rock classification was based upon the mode of formation, the geological age, and the specific mineral properties, which
includes the nature of the mineral and the structure of
the rock.

Subsequent editions of the original works of Rosenbusch and Zirkel, and a number of new noteworthy contributions by Roth (1883), Teall (1888), Loewinson-
Lessing (1890-1897), and Johannes Walther (1897)
appeared, chief attention being given to the classification
of igneous rocks on the basis of origin, age, and characters.

Within the last twenty years a number of American
petrographers have made noteworthy contributions to
the science of rock classification, and with the coöpera-
tion of the field geologist who has gradually become more
and more painstaking in the matter of collecting and
labeling rock specimens for future study, they hope to
evolve from the present classification which is marred
by a complexity of nomenclature, a logical and compre-
hensive system of classification which will approach in
construction as closely as possible a truly natural arrangement.

Among the earlier American petrographers who made
valuable contributions toward the development of the
science are J. F. Kemp, J. S. Diller, Whitman Cross, J. P.
Iddings, and F. D. Adams.

PART ONE. OPTICAL MINERALOGY.

CHAPTER 1.

The Elements of Optics, and the Application of Polarized Light to Crystalline Substances.

The Nature of Light.—Light is a form of energy which in a homogeneous medium as the ether is transmitted in a rapid wave motion in straight lines with no change in the direction of propagation. This wave motion is considered to be a resultant of simple harmonic motion and a uniform motion at right angles to this. In other words, wave motion is a vibration which takes place at right angles to the direction of propagation of the light.

A ray of light is a line which designates the direction of transmission of the wave. The intensity of light depends upon the rate or wave-length of the vibrations. Color sensation is determined by the number of waves of light which reach the eye in a given time. The wave-length for red light is 760 millionths of a millimeter, and the wave-length for violet light is 397 millionths of a millimeter. White light is the sum of light of all these various wave-lengths. The velocity of light of all colors in vacuo is the same, and is about 300,000 km per second.

Isotropic Media.—Light is transmitted with equal velocity in all directions in certain media, as air, water, and glass. Light which is transmitted through such a medium if it finds its source in that medium will be propagated as spherical waves, in which the wave-front or wave-surface forms a continuous surface, and all points on that surface are equidistant from the source. A ray of light is perpendicular to its wave-front.

In an isotropic substance, this wave-surface may be represented by the surface of a sphere. Any plane passing through this imaginary sphere in any position will have a circular outline. Gases, liquids, amorphous substances as volcanic glass, and crystals of the isometric system, are isotropic substances. The velocity of transmission of light through these substances is independent of the direction of vibration.

Anisotropic Media.—In anisotropic media (as opposed to isotropic media), the velocity with which light is propagated varies with the direction. All substances which are not amorphous or which do not belong to the isometric system are optically anisotropic.

Anisotropic crystals are divided into uniaxial and biaxial crystals.

Uniaxial Crystals.—In uniaxial crystals, only one direction exists in which there is no double refraction of light. This is in the direction of the vertical crystallographic axis, which is called the optic axis. It lies in the direction of either the greatest or least ease of vibration. The wave-front which represents the optical structure of uniaxial crystals is an imaginary spheroid of revolution in which the optic axis is the axis of revolution. A plane passing through the spheroid in any direction at right angles to the optic axis has a circular outline. Any other section has an elliptical outline. Tetragonal and hexagonal crystals are uniaxial.

Biaxial Crystals.—In biaxial crystals there are two directions corresponding in character to the one optic axis of uniaxial crystals, which gives rise to the term biaxial. The wave-front which represents the optical structure of biaxial crystals is an imaginary ellipsoid with three unequal rectangular axes. A plane passing through this ellipsoid in any direction at right angles to either of the optic axes has a circular outline. Any other section has an elliptical outline. Orthorhombic, monoclinic and triclinic crystals are biaxial.

Index of Refraction.—The previous discussion has been concerned with light which has. passed through homogeneous media. If a system of light waves of the same wave-length passes obliquely from one medium into another, there will be a change in the direction of transmission depending upon the relative ease or difficulty with which the light may penetrate the new medium. If the second medium, such as glass, is optically more dense than the first medium, such as air, that portion of the wave-front which first strikes the glass will experience a greater difficulty in transmission, and its velocity will be reduced, while the remainder of the wave-front is still traveling with the same velocity in the air. When this portion of the wave-front finally reaches the glass, it has gained upon the first portion, with a result that the wave will have suffered a deflection from its original course. From this position the various portions of the wave-front continue through the glass with equal velocities.

This phenomenon is called refraction. It is a change of direction at the bounding surface. Refraction is toward the perpendicular (to the bounding surface) when the passage of a light ray is from the rarer to the denser medium, and away from the perpendicular in the opposite case.

In Fig. 1, D C is the bounding surface between two media, of which the lower is optically denser than the upper. G H is a perpendicular to the bounding surface. Angle i is the angle of incidence and angle p is the angle of refraction. A constant relation exists between the sines of these angles regardless of the direction of transmission, which may be expressed as follows: the sine of the angle of incidence bears a constant ratio to the sine of the angle of refraction. This ratio may be expressed by the equation $\dfrac{\sin i}{\sin r} = n$, in which n is the index of refraction and is inversely proportional to

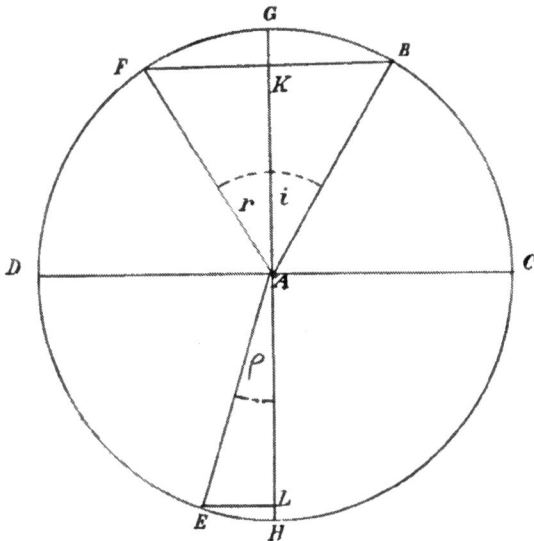

Fig. 1.   Reflection and single refraction.
(Winchell.)

the wave velocity. In this formula there are two limiting relations to be considered. If $i = 0$, $r = 0$, in which case the angle of refraction becomes zero. Thus, by perpendicular incidence, the ray proceeds in the second medium without any change in direction. If $i = 90$, $\dfrac{1}{\sin r} = n$ or $\sin r = \dfrac{i}{n}$. This value of r is known as the critical angle, or angle of total reflection, and may be defined as that angle beyond which no light passes from denser to rarer media. All light may pass from a rarer to a denser medium, but the amount of light which may pass from the denser to the rarer medium is limited by the critical angle. The critical angle is a constant for the substance.

Thus for water $n = 1.335$.

$$\sin r = \frac{1}{1.335.}$$

$$r = 48° \ 35'.$$

And for diamond, $n = 2.42$.

$$\sin r = \frac{1}{2.42}$$

$$r = 24° \ 25'.$$

If light is to pass from water into air, the rays must strike the surface at angles less than 48° 35', whereas if light is to pass from diamond into air, the rays must strike the surface at angles less than 24° 25'. Evidently more light can enter the diamond than can directly escape, and this fact is responsible for the brilliancy of the gem. The greater the index of refraction, the smaller will be the critical angle. In a cut diamond, the facets are arranged so that most of the light is totally reflected.

Most substances have a value for n ranging between 1 and 2. The following table gives the indices of refraction for a variety of substances:

Ice	1.310	Quartz	1.547
Water	1.335	Calcite	1.601
Alcohol	1.36	Methylene iodide	1.75
Fluorite	1.434	Garnet	1.814
Common glass	1.435	Zircon	1.952
Olive oil	1.47	Sphalerite	2.369
Canada balsam	1.536-1.549	Diamond	2.429
Rock salt	1.544	Rutile	2.712
Bromoform	1.59	Pyrargyrite	3.016

An adamantine luster is characteristic of minerals with an index of refraction above 1.9.

Ordinarily, the index of refraction of a substance is determined by passing the incident ray into the substance from air, but other media than air might be used. The index of refraction of the substance in air is the product of its index in the medium by the index of that medium in air. The index of refraction òf air when referred to a vacuum is 1.000294.

Elementary phenomena in refraction, such as the apparent bending of a stick of wood when partially submerged in water, were no doubt observed in early times. The constant ratio between sin i and sin r was first established by Descartes in 1637, but it was not until Newton succeeded in producing a colored spectrum by a prismatic decomposition of white light that the full importance of n was realized.

Double Refraction.—Double refraction is the property possessed by all anisotropic crystals of resolving a light ray into two rays polarized at right angles to each other and traveling in different directions. This is due to the fact that upon entering the anisotropic medium the vibrations of light are made to conform to the molecular structure of the medium. In other words, light travels with different velocities in different crystallographic directions in the same substance.

The ray advances with the greatest velocity when it is vibrating parallel to the direction of the greatest ease of vibration, and with least velocity when vibrating parallel to the direction of least ease of vibration. These rays obviously have different indices of refraction. The ray which follows the usual laws of single refraction is called the ordinary ray, expressed by O. The other ray is called the extraordinary ray (E) be-

cause it does not follow the usual laws of single refraction.

When a ray of light enters an anisotropic medium
perpendicular to its surface, the ordinary ray passes
through without suffering refraction, provided the sur-

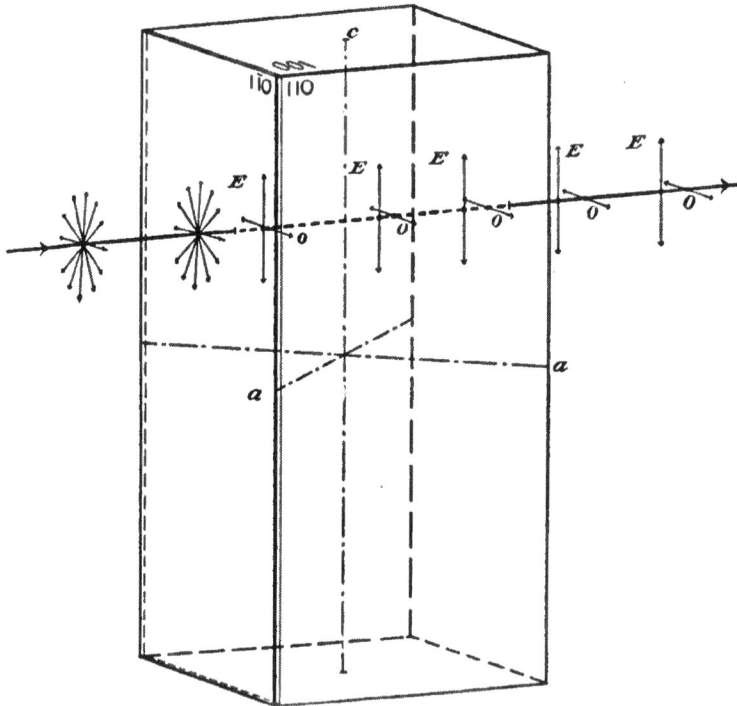

Fig. 2. Plane wave advancing perpendicular
to the vertical axis, showing ether
vibration and retardation of
the O and E rays.
(Winchell.)

face through which it emerges is parallel to the surface through which it enters. The extraordinary ray
is diverted. To this rule the following exceptions must
be noted. If the original ray enters a substance perpendicular to the surface and at the same time par-

allel to an optic axis, there is no refraction nor polarization. If perpendicular to the surface and to an optic axis, there is no refraction but there is a division into two rays traveling with different velocities and polarized at right angles to each other.

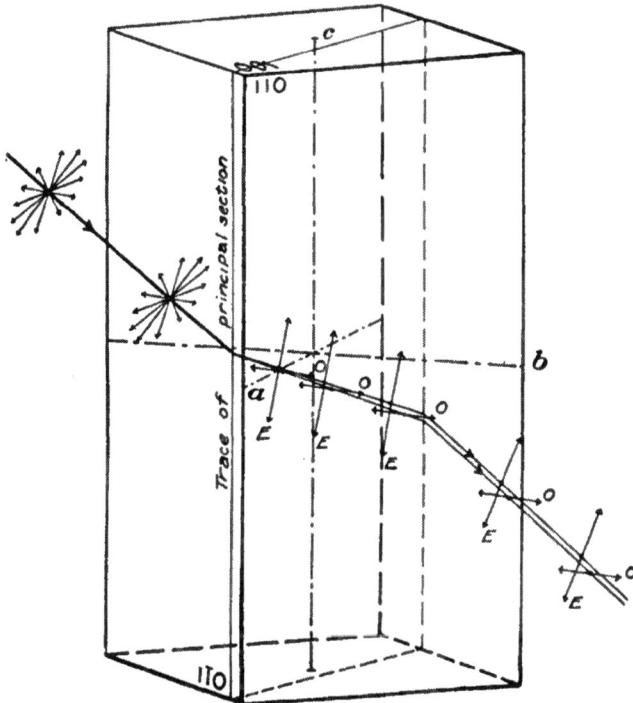

Fig. 3.   Oblique incidence on a surface
parallel to the optic axis.   (Winchell.)

When the two rays emerge from the substance, they resume parallelism, but the waves of one are slightly in advance of the waves of the other. Such waves are said to interfere with each other, producing light of different colors. Upon this phenomenon is based much of the work done in examining minerals in thin section under the microscope in parallel polarized light.

Interference.—Two waves of like length and amplitude, if propagated in the same direction and meeting in the same phase, unite to form a wave of double amplitude. If these waves differ in phase by half a wavelength or an odd multiple of this, they interfere in such a way as to extinguish each other. For other relations of phase falling between these extreme cases they also interfere with each other, forming a new resultant wave, differing in amplitude from each of the component waves. We are assuming here the use of monochromatic light waves, or light waves of like length. If ordinary white light is employed, the waves in case of interference will be indicated by the appearance of the colors of the spectrum.

Polarization.—Ordinary light is propagated by transverse vibrations of the ether which take place in all directions about the line of propagation. Plane polarized light is propagated by ether vibrations which take place in one plane only. This phenomenon is called polarization. It may be described as a change in the character of reflected or transmitted light, which diminishes its power of being further reflected or transmitted.

Light is polarized by reflection, by single refraction, and by double refraction. The plane of polarization of light polarized by reflection is defined as the plane containing the incident and the reflected rays, the vibrations taking place at right angles to it. The plane of polarization of the refracted ray is the plane at right angles to the vibration direction, consequently at right angles to the plane of the incident and the reflected rays. That light is polarized when reflected may be shown experimentally by the use of two reflecting surfaces.

Nicol Prism.—The Nicol prism, so named after its

inventor, Nicol, is a device for producing polarized light. It consists of a clear transparent crystal of calcite known as Iceland spar, as it is obtained almost exclusively from caves in certain basalts in Iceland. The vertical faces are natural cleavage faces, in which the end cleavages, inclined 71 degrees to the obtuse edges of the prism, are ground down and polished so as to make an angle of 68 degrees with the obtuse vertical edges. It is then cut diagonally in two parts perpendicular to

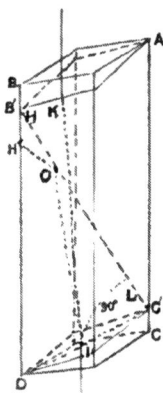

Fig. 4. Side view of
the Nicol prism.

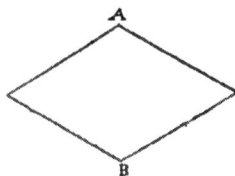

Fig. 5. End view of
the Nicol prism.

the short diagonal of the end face. The two parts are cemented together in their original position by Canada balsam, a resin obtained from a species of fir. It has an index of refraction of 1.536. Since calcite is a doubly refracting substance, the Nicol prism refracts a ray of light into two rays — the ordinary ray, having an index of 1.658, and the extraordinary ray, having an index of 1.486.

The angle at which the two new planes are polished, as well as the angle at which the crystal is cut, are so calculated that the ordinary ray will strike the balsam at an angle greater than the critical angle. Consequently, the ordinary ray is totally reflected and is absorbed in the blackened walls of the cork mountings. The extraordinary ray passes through the balsam as a completely polarized ray which is vibrating in a known direction, namely, parallel to the short diagonal of the calcite rhomb. When two Nicol prisms have their short diagonals parallel, light passes through without being changed except for a decrease in intensity. If one of the nicols is revolved, the light gradually diminishes until the nicols are at 90 degrees to each other, when darkness results.

CHAPTER 2.

The Polarizing Microscope and Its Parts.

The Polarizing Microscope.—In order to ascertain the peculiarities of minerals of each of the crystallographic systems as they are manifested in polarized light, the polarizing microscope is used. This instrument is applicable to the study of the form, optical properties, and mutual relations of the minerals as they are found in thin sections of rocks, making it a valuable aid to geological research. It is likewise used to great advantage in the study of small isolated crystals, or fragments of crystals. A determination of the following characteristics of the unknown mineral is of particular value in its identification: crystal form as shown in outline, direction of cleavage lines, refractive index, light absorption in different directions, the isotropic or anisotropic character, position of the axial plane and the nature of the axial interference figures, the strength and character (positive or negative) of the double refraction, presence and nature of inclusions, type of twinning!

In addition to the parts essential to the ordinary microscope the polarizing microscope contains the following parts: two Nicol prisms, a lens for convergent polarized light, a rotating stage, and an ocular with cross hairs.

Nicol Prisms.—The effects due to polarized light cannot usually be distinguished except by a combination of two Nicol prisms. The upper nicol is not revolvable,

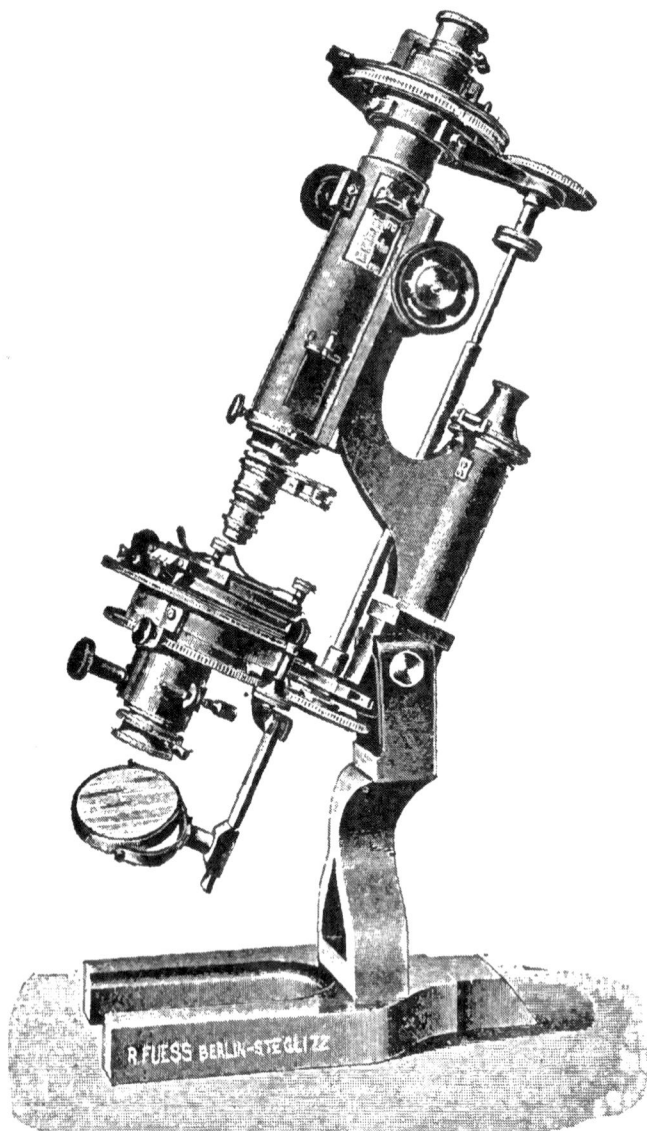

Fig. 6. The Fuess microscope.

and is placed in a support between the ocular and the objective. It can be pushed in or out of the tube at will. It is called the analyzer.

The lower nicol, which is revolvable, is placed beneath the stage. For ordinary work, its principal section (i.e., its shorter diameter) is placed at right angles with the principal section of the upper nicol. It may be raised or lowered without disturbing the centering. This nicol is called the polarizer.

The principal section of the analyzer is left and right, and that of the polarizer is front and rear. In this position the field is dark and the nicols are "crossed."

When a thin section is examined over the lower nicol or between two nicols without the convergent light, it is said to be done in parallel polarized light. When the lower nicol is used alone, its vibration plane must be known. A simple test is to place a cleavage fragment of calcite within the field of view. It has a high relief when its long diagonal is parallel to the plane of vibration of the nicol.

A section of biotite cut at right angles to its cleavage has its greatest absorption when its cleavage direction is parallel to the plane of vibration of the polarizer. Consequently, it is darkest in this position. Tourmaline, on the other hand, extinguishes vibrations at right angles to the optic axes, i.e., it absorbs the ordinary ray, and only the light rays vibrating parallel to the crystallographic axis c emerge.

By removing the Nicol prism from the tube, the separating plane of the balsam along which the two fragments of calcite were cut may be seen upon looking through the prism at an angle. The vibration direction of the ray which passes through the prism (the extraordinary ray) is normal to the layer of balsam.

The polarizer may be removed from the microscope

and light reflected from a horizontal surface, such as a plate of glass or a table top, examined through it. Since light is polarized in a plane parallel to the reflecting surface, the polarizing plane of the nicol lies at right angles to the reflecting surface when the latter appears dark.

Lens for Convergent Light.—When the operation demands convergent light, a powerful convergent lens can be thrown into the tube of the microscope over the polarizer by means of a lever beneath the stage. This lens may be raised with the lower nicol until the surface of the lens is practically in contact with the glass slide holding the thin section.

The Rotating Stage.—The stage is a circular table upon which the thin section is placed for examination. The edge has a graduated scale and vernier reading to minutes. The center of the stage must coincide with the optical center of the tube. Centering is done by means of two centering screws, 90 degrees apart, located on the lower end of the tube. The thin section is held in place by two spring object clips. Recent forms of microscopes are equipped with mechanical stages which have freedom of movement in a left-and-right direction and in a front-and-rear direction, thus allowing a rapid inspection of every part of the section.

Cross Hair.—Cross hairs are placed in the ocular at right angles to each other, one running left and right and the other front and rear, in agreement with the principal sections of the nicols when crossed.

To measure a plane angle in thin section, or the interfacial angle of a small, flat crystal, the stage is centered with the intersection of the two edges at the center of the cross hairs. A reading is made when one edge of the crystal is parallel to the left-and-right cross hair, then the stage is revolved until the other edge is parallel to the same cross hair but on the opposite side of the center. Another reading is taken. The difference between the two readings is the external angle.

Other parts of the microscope which deserve explanation and suggestions as to proper use are taken up in order.

The Mirror.—The mirror which is attached to the substage reflects light from the source to the object. A plane mirror forms one side and a concave mirror the other. The former is used for low magnification, where a weak light is sufficient. The latter is used for higher magnifications. This mirror concentrates the light by converging the rays included within an angular aperture of about 40 degrees. For still higher magnifications and for all phenomena observed in convergent light the condensing lens is used.

A proper use of the mirror is essential to the most efficient use of the microscope. When parallel rays, such as ordinary daylight, are used, they are reflected from the mirror with a slight loss of intensity. They are reflected from the concave mirror with increased intensity, the rays coming together at the focal point. If the source of light is close to the instrument, the focal length is larger. To meet this adjustment, the mirror is attached to a sliding vertical bar. Since the condensing lens has its focus some distance above its upper surface, the plane mirror is used in connection with it.

The Objective.—Objectives are classified according to their magnification. An objective of low power has a focal length above 13 mm and a magnification less than 15 diameters; it is of medium power when its focal length is between 12 and 5 mm and its magnification is 40 diameters; of high power when its focal length is less than 4.5 mm and its magnification exceeds 40 diameters. The objectives most commonly used are numbers 3 and 7, the former for searching out an object and for making the preliminary examination, and the latter for convergent light and high power.

A thin section may be considered as made up of a series of planes superimposed one above the other, only one of which may be seen for one adjustment of focus. With low-power objectives one can see objects lying in slightly different planes, but with high-power lenses this is impossible, as the depth of focus diminishes inversely as the numerical aperture. The brightness of the image increases as the square of the numerical aperture.

Resolving Power.—The resolving power of an objective is that property by virtue of which one is able to see the finer details of an object. This resolving power increases with the number and obliquity of the rays coming from the object, consequently an immersion fluid by increasing the number of rays brought to the object increases the resolving power. In petrographic work no very great magnifying powers are required, and immersion lenses are not much used except for particular kinds of work.

When two points are removed from the eye 6,876 times the distance separating them they will appear as a single point. The eye is able to distinguish only about 250 lines to an inch. Thus pleurosigma angulatum with about 50,000 lines to the inch can be resolved by a one-half inch objective so as to be clearly seen with a three-quarters inch ocular but not with one-and-one-half inch. A much smaller line may be seen than the interval between two lines.

Cost of Objectives.—Objectives with a focal length of 25 mm and over cost about $4 each; between 25 and 10 mm, $5.50 to $10; 10 to 3 mm, $7 to $15; 3 to 2 mm, about $20. Students are urged to treat them with care.

Bertrand Lens.—In the center of the microscope tube above the analyzer is the Bertrand lens, which may be thrown in or out of the tube by means of a sliding carrier. It acts as a small microscope which is used with the ocular to magnify interference figures.

The Ocular.—The Huygens ocular which is most generally used in petrographic microscopes consists of two simple plane-convex lenses placed with their plane surfaces toward the eye. The upper lens is known as the eye lens and the lower as the collective or field lens. The focal length of the eye lens is one third of the field lens, and they are separated a distance equal to the sum of their focal lengths. The rays of light emerging from the eye lens are parallel and thus cause the eye less fatigue.

The cross hairs which are placed in the eyepiece are made of spider web, ·the dark thread from the inside of the nest being the best.

Micrometer.—It is desirable at times to measure small distances such as the dimensions of small crystals. A special eyepiece called the micrometer has been devised ·for this purpose. It contains a scale etched on glass. On the stage of the microscope a scale reading to hundredths of a mm is placed. It is then necessary to find to how many hundredths of a mm each division of the eyepiece is equivalent.

It may be well for the student, in order to become familiar with the use of the micrometer, to construct a table showing the value of the ocular micrometer for each objective. The stage micrometer is used for this purpose.   ·

Adjustment Screws.—The tube carrying the eyepiece and objective has a fine adjustment screw, the edge of which is graduated. It moves against a fixed index attached to the tube, by which means the distance through which the tube is raised or lowered can be measured to .001 mm.

The student is advised as a laboratory illustration to determine the amount which one revolution of the fine adjustment screw raises the objective. To do this, measure the thickness of a glass plate or cover glass by focusing carefully on the lower surface of the glass and then upon the upper surface. This distance is measured by the micrometer by setting the glass plate on edge, slightly embedded in paraffin or wax, or supported otherwise.

Use of the Microscope.—The best light for microscopic work is that coming from the north; the next best from the east. Direct sunlight should never be used. The table should be firm, and of a height to suit the convenience of the individual. The instrument should be placed directly in front of the observer, so that both hands can be used for manipulation.

The eye which is not used for observation should also be kept open. Although it may seem difficult at first to concentrate the gaze on the thin section, it will be found to be far less fatiguing. When using high powers, the eye must be kept very close to the ocular, with low powers slightly farther removed. When both nicols are being used, more light is advantageous than when only one is in use. As much of the examination of a thin section as possible should be done with the low powers in order to save a strain on the eyes.

The student is particularly cautioned whenever focusing with high powers to focus upward and never downward. If this rule is followed, no thin sections will be broken. Place the eye on a level with the stage, and lower the tube slowly until the objective is almost in contact with the thin section. Then looking through the tube, raise the objective slowly until the portion of the section desired for examination is in focus.   If colorless minerals such as quartz are being examined, it is well to reduce the amount of the illumination and look for bubbles or other inclusions.

Proper care of the nicols and lenses prolongs their life and increases their efficiency.   They should not be exposed to severe sunlight nor to the heat from a steam radiator, lest the cement soften.   The lenses should be kept free from dust.   The objective should never be allowed to come in contact with the cover glass.

CHAPTER 3.

General Methods of Mineral Determination.

The determination of unknown minerals in thin section may be accomplished by the use of one or all of the following eight general operations:

1. Determination of the general physical properties of minerals by ordinary light.

2. Determination of the relative refractive index.

3. Determination of the relative double refraction or birefringence.

4. Determination of the axial interference figures.

5. Determination of the dispersion of the optic axes.

6. Determination of the optical character or the character of the double refraction.

7. Determination of the extinction angle or the relation of the crystallographic axes to the axes of ether elasticity.

8. Determination of the presence or absence of pleochroism.

General Operation No. 1: Determination of the General Physical Properties of Minerals by Means of Ordinary Light.

The physical properties of minerals referred to in this paragraph are: crystal form, cleavage, parting, twinning, and color. Ordinary light is light which has not been polarized to obtain which both nicols should be removed from the microscope. If the lower nicol is difficult to remove, the observations are made in plane polarized light, which generally causes no great difference in the appearance of the mineral. The intensity of the unpolarized light, however, is much greater than that of the polarized.

Minerals examined by ordinary light are of two classes: transparent and opaque. The former class is examined by transmitted light for crystal form, cleavage, and color; the latter class by incident light for crystal form, color, luster, etc.

CRYSTAL FORM. The determination of crystal form is not of great importance in the study of rock sections for the reason that individual crystals have not had the opportunity for undisturbed development, but have been hampered in their growth by interference with neighboring crystals. In certain porphyries a study of the form of the phenocrysts often leads to their identification.

CLEAVAGE. Pronounced cleavage lines are developed in certain minerals in characteristic directions during the process of grinding to thin section. The direction and perfection of the cleavage cracks is indicative. A mineral possessing no cleavage will have irregular cracks, as quartz.

Perfect cleavage is a cleavage in which the lines are sharp and extend for considerable distances. Examples: mica, fluorite.

Good or distinct cleavage is cleavage in which the cracks are interrupted with offsets, etc. Examples: augite, hornblende, orthoclase.

Poor or indistinct cleavage is very irregular, with uneven cracks, though they follow roughly certain directions.

Pinacoidal cleavage, as shown in mica, is well developed in one direction only. Prismatic cleavage, as shown by augite and hornblende, usually develops in two planes. In certain minerals of the isometric and hexagonal systems, such as galena and calcite, three good cleavages develop.

Cleavage angles, of course, depend upon the orientation of the random section shown in the thin section. Where the section is cut at right angles to the cleavage planes, the angles are characteristic. Hence, if one is using cleavage fragments, he can orient it at will, since the flat faces will bear definite relations to the crystallographic axes.

PARTING. Parting is a fracture often developed parallel to a certain cleavage direction occurring along planes of weakness as may result from shearing or gliding planes.

TWINNING. Twinning is important in certain minerals and will be discussed in Chapters 5, 6, and 7.

COLOR. All colored minerals may be divided into two classes: idiochromatic and allochromatic. Idiochromatic minerals are those in which the color is due to a property of the mineral itself, namely, its ability to absorb light of certain wave-lengths, although the property of absorption may not be the same in every direction. Allochromatic minerals are those in which the color is due to inclusions, which may or may not be distinguished under the microscope. The pigment may be either organic or inorganic. Carbon, nitrogen and hydrogen have been found in zircon, smoky quartz, amethyst, fluorite, apatite, calcite, microcline, barite, halite and topaz. Free fluorine has been found in fluorite. Traces of iron are found in brown zircon. The pigment may be thickly and evenly distributed or irregularly and so sparingly distributed that a thin section appears colorless.

General Operation No. 2: Determination of the Relative Refractive Index.

Refraction is the change which light undergoes in direction in passing between two media which differ in density. The index of refraction described under optics may often be judged by the appearance of the mineral in the liquid in which it is mounted, usually Canada balsam. This makes a convenient standard with which to compare the index of refraction of the unknown mineral. Although its index varies slightly during the process of mounting and with age, only those few minerals whose indices fall within the limits of variation of the balsam are affected. Balsam may retain its sticky consistency and low index for forty years if protected by a cover glass.

If the mineral under examination and the balsam have practically the same index of refraction, the mineral will appear smooth and will be visible with difficulty. It is then said to have "low relief." If the two have quite different indices, the surface of the mineral will appear rough and the borders dark. Such a surface, because of its resemblance to shagreen, is called a shagreen surface. The mineral is said to have "high relief." This apparently rough surface is due to inequalities of the surface, each elevation and depression reflecting and refracting the light at a different angle. This irregular illumination causes the mineral to appear darker in some spots and lighter in others. A mineral embedded in Canada balsam will have high relief whether its index is lower or higher than the liquid.

In a rock section where a number of different minerals having different indices of refraction lie in contact, certain minerals appear to stand out above the others in relief. Minerals with high indices seem to be elevated from the plane of the section. This is because the rays of. light from the lower surface of different minerals appear to come from the points of intersection of the refracted rays.

Since the index of refraction of a mineral is one of its most important optical properties, many methods have been devised for its identification.

THE METHOD OF DUC DE CHAULNES. By this method one may determine the index of the mineral directly by focusing a medium- or high-power objective accurately upon an object, and then inserting between it and the objective a transparent plate with parallel sides. The image becomes blurred. The tube of the microscope is raised until the image is again in focus. The amount of change necessary is dependent upon the index of refraction of the plate and upon its thickness.

As a laboratory illustration the student is advised to determine the index of refraction of a plate glass or cover glass by this method. If the student uses a glass whose true thickness has already been determined, the index may be obtained by measuring the apparent thickness and dividing the true thickness by this latter amount. If the thickness of the glass is not known, it is possible to determine the approximate true thickness by focusing on a point or scratch on another plate of glass which is to be used for a support. Then place upon it the glass plate whose thickness is to be determined, and focus on its upper surface. This distance is the thickness of the glass plate plus the thickness of the air film separating the plate from the support, and should be used only in case the thickness of the plate is so great that the thickness of the air film becomes negligible. The apparent thickness can now be measured and the index calculated as mentioned above.

A correction for the air film can be made very easily and should always be done if the glass plate is thin.

Focus on the upper surface of the glass plate and then on the lower surface. This distance through which the objective moves is the apparent thickness of the plate. Now focus on the surface of the support. This gives the true thickness of the plate and the air film. Next focus on the lower surface of the plate and on the upper surface of the support. This gives the true thickness of the air film, which can readily be subtracted from the thickness of the glass plate and air film. The difference is the true thickness of the glass plate. The index can now be determined by dividing the true thickness by the apparent thickness of the plate.

IMMERSION METHOD. If a drop of a liquid with an index equal to that of the mineral is placed upon a thin section of the mineral without a cover glass, the appearance of roughness which characterizes the mineral in air disappears, since there is neither reflection nor refraction at the contact, and the light passes through without deflection. If the mineral is colorless, it practically disappears from view. By the use of a series of immersion liquids whose indices of refraction are known, it is possible to experiment with the unknown mineral until a liquid is found whose index of refraction by the above test corresponds with the index of refraction of the mineral.

BECKE METHOD. By the Becke method, which involves the use of total reflection in connection with refraction, one may determine the relation which the refractive index of the unknown mineral has to that of one which is known and which is in contact with it. Bring the focus directly upon the line of separation of the two minerals, using a high-power objective in convergent light. If the condenser is lowered and the analyzing nicol is removed, it is observed that the field becomes slightly darker, and a fine line of white light sharply marks the contact of the two minerals. Upon raising the objective very slightly, this thin line of white light will be seen to shift from the line of contact of the two minerals toward the mineral having the higher index.

This phenomenon is explained as follows: The rays of light which enter the minerals perpendicular to their surfaces undergo no refraction but pass directly through. Those rays which enter the mineral having the lower index of refraction reach the plane of contact of the two minerals and are all bent toward a normal to this plane passing through the mineral having the higher index, because they are passing from a rarer to a denser medium. The rays which enter the mineral having the higher index must pass from a denser medium to a rarer. In such a case it is remembered that all of those rays which strike the mineral of lower index at an angle greater than the critical angle of the denser mineral, are totally reflected and emerge from the upper surface of the denser mineral. Only those rays pass into the mineral of lower index which strike the plane of contact of the two minerals at an angle less than the critical angle of the mineral of higher index.

Upon lowering the objective slightly, the white line shifts toward the mineral with the lower index.

It is advisable to use the Becke test on contacts which are nearly or quite vertical. One can easily determine a vertical contact by shifting the focus and seeing that the boundary remains sharp at all foci, and in the same position. The verticality of the contact makes no difference with the result, provided the medium having the lower index lies above. If it lies below and the inclination is great enough, the bright line may appear to move the wrong way.

According to this method, differences of .001 between indices are noticeable. It is especially useful in determining minerals with low indices, as sodalite, leucite, or in distinguishing between orthoclase and quartz. Since the mean refractive index of quartz is about the same as andesine, is higher than orthoclase, albite and oligoclase, and less than labradorite, bytownite and anorthite, certain definite inferences may be drawn regarding these minerals in contact.

The following scale of refringence (after Winchell) will aid the student to estimate the value of the mean index of refraction of minerals in thin section by means of "reliefs."

SCALE OF REFRINGENCE.

Very low refringence.	Example, fluorite, $n = 1.434$.
Low refringence.	Example, quartz, $n = 1.547$.
Moderate refringence.	Example, hornblende, $n = 1.642$.
High refringence.	Example, augite, $n = 1.715$.
Very high refringence.	Example, zircon, $n = 1.952$.

The "negative" relief seen in fluorite is caused by the total reflection of light striking the lower surface of the mineral.

The indices of an unknown mineral compared with those of any known mineral by the Becke method will always give at least one limit, which in connection with the visible amount of relief may be sufficient.

General Operation No. 3: Determination of the Relative Double Refraction or Birefringence.

ISOTROPIC CRYSTALLINE SUBSTANCES. — ISOMETRIC MINERALS. Between crossed nicols, isometric minerals remain dark in thin section during the entire revolution of the stage. Such minerals allow the rays to vibrate with equal ease in all directions regardless of the direction in which the section is cut. They have no interference colors in parallel polarized light nor interference figures in convergent light.

This is due to the fact that, in isotropic substances, light is transmitted with equal velocities in all directions; hence the velocity of light transmission is independent of the direction of vibration.

Certain important isometric minerals, as pyrite and magnetite, are opaque. These minerals are examined and identified by other means than by polarized light.

ANISOTROPIC SUBSTANCES. Uniaxial minerals, or minerals of the tetragonal and hexagonal systems.

A ray of light emerging from the polarizer (lower nicol) is vibrating in one plane only, left and right. Upon entering a thin section of a tetragonal or hexagonal mineral cut perpendicular to the vertical crystallographic axis c, the light is not disturbed, because about this axis vibrations take place with equal ease in all directions perpendicular to it. In this one direction, light is singly refracting.

The *optic axis* is that direction in a doubly refracting substance in which light is singly refracting. Hence in uniaxial minerals the crystallographic axis c coincides with the optic axis.

This ray of light emerging from the mineral is intercepted by the analyzer so as to produce darkness. The student should take the precaution, upon observing a mineral which remains dark throughout a complete revolution of the stage, to determine whether it is an isometric mineral or an isotropic mineral cut perpendicular to an optic axis.

A ray of light from the polarizer entering the thin section of an anisotropic mineral in any other direction than perpendicular to the optic axis is doubly refracted and polarized, the extraordinary and the ordinary rays advancing in different directions, the former taking the oblique direction. These rays are, of course, vibrating perpendicular to each other, and perpendicular to their direction of propagation. The extraordinary ray vibrates in a plane containing the incident ray and optic axis. This plane is called the *principal optic section*.

On emerging from the thin section, the extraordinary ray is more or less advanced than the ordinary ray. Upon reaching the analyzer, each of these rays is again resolved into two rays — an extraordinary and an ordinary — the two extraordinary rays vibrating in one plane, and the two ordinary rays in a plane at right angles. Upon reaching the layer of Canada balsam, the ordinary rays are totally reflected and absorbed. The two extraordinary rays emerge from the analyzer in a uniform direction, but not equally advanced, consequently in different phases. This interference produces color. If the rays have a difference of phase of one half of a wave-length, or any uneven multiple thereof, darkness will result.

COLOR. The kind of color produced depends upon:
1. The mineral.
2. The thickness of the section.
3. The direction in which the section is cut.

The amount of color depends upon the angle between the principal optic section and the principal section of either nicol. The color is least when the angle is 0 degrees, and greatest when the angle is 45 degrees, each of these conditions occurring four times in one revolution of the section. Upon this phenomenon is based the determination of the angle of extinction (General Operation No. 6).

AXES OF ETHER VIBRATION. The direction in which ether vibrates in anisotropic minerals with the greatest ease is called the greatest axis of ether vibration. It

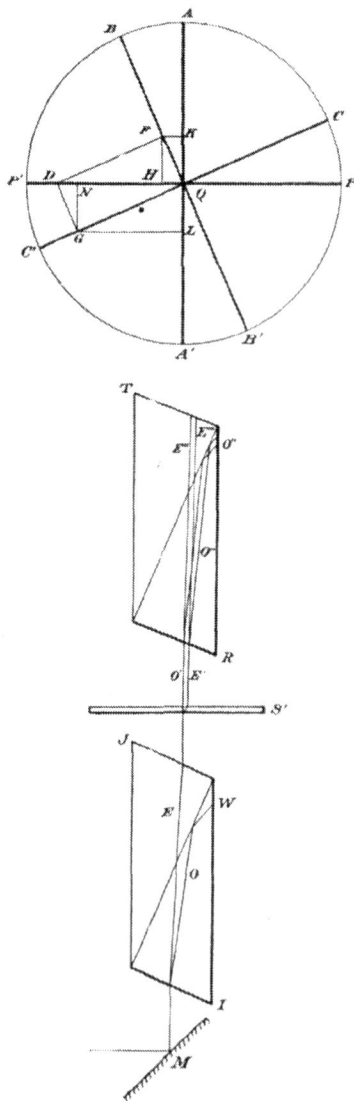

Fig. 7.   Changes of light in passing through
a petrographical microscope.

is denoted by the letter X.   The index of refraction of light vibrating in this direction is expressed by n_p.

The direction in which ether vibrates in anisotropic minerals with the least ease is called the least axis of ether vibration.   It is denoted by the letter Z.   The index of refraction of light vibrating in this direction is expressed by n_g.

In uniaxial minerals, one of these axes always coincides with the vertical crystallographic axis c.   The other axis is in all directions at right angles to this. Either the greater or the lesser axis of ether vibration may coincide with the vertical crystallographic axis.

The value of the maximum double refraction or birefringence is the difference between n_p and n_g.   Thus, for calcite, n_g is 1.658 and n_p is 1.486; $n_g - n_p = 0.172$, which indicates a very strong birefringence.   For quartz, n_g is 1.553 and n_p is 1.544; $n_g - n_p = 0.009$, which indicates weak birefringence, as the retardation of one ray over the other in emerging is very slight.

It will be observed that in certain minerals the index of refraction of the extraordinary ray is greater than that of the ordinary ray, and in other minerals the reverse is true.   A further discussion of this fact will be taken up under General Operation No. 5.

NEWTON'S COLOR SCALE.   Thin sections of anisotropic minerals cut not perpendicular to an optic axis show polarization colors between crossed nicols.   A careful study of these colors is most important for a successful determination of unknown minerals.

The color scale of Newton has been adopted as a standard.   It consists of a succession of interference tints shading into each other.   These same tints are produced by anisotropic minerals in thin section.   About forty distinguishable tints in natural light have been

named. The colors are best exhibited by thin sections
which have a thickness ranging between 0.01 and 0.06.

Newton's color scale as applied to the principal rock-
making minerals in sections 0.03 mm thick is given here-
with.

NEWTON'S COLOR SCALE. ·

Millionths of a mm Retardation	Interference of Colors Between X Nicols.	$n_g - n_p$	Rock-forming Minerals.
0	Black		
30	Iron gray.	0.001	Leucite.
60	Lavender gray.	0.002	Vesuvianite.
117	Lavender gray.	0.004	Apatite.
140	Grayish blue.	0.005	Beryl.
153	Grayish blue. ·	0.005	Nephelite, riebeckite.
180	Lighter gray.	0.006	Stilbite, zoisite.
218	Lighter gray.	0.007	Orthoclase, microcline, kaolin.
234	Greenish white.	0.008	Oligoclase, albite, labradorite.
250	White.	0.009	Corundum.
267	Yellowish white.	0.009	Gypsum, enstatite.
281	Straw yellow.	0.009	Quartz, sapphire.
306	Light yellow.	0.010	Topaz, rhodonite, staurolite.
332	Bright yellow.	0.011	Clinochlore, barite. Andalusite.
390	Orange yellow.	0.013	Anorthite, hypersthene.
433	Orange yellow.	0.014	Wollastonite.
474	Orange red.	0.016	Cyanite.
575	Violet (Sensitive tint No. 1)—Green, Yel.	0.019	Hedenbergite.
589	Indigo.	0.020	Tourmaline.
629	Blue.	0.021	Wernerite.
667	Sky blue.	0.022	Augite.
688	Sky blue.	0.023	Hornblende.
713	Greenish blue.	0.024	Diallage.
747	Green.	0.025	Actinolite. Augite.
810	Light green.	0.027	Tremolite, arfvedsonite.
855	Yellowish green.	0.029	Diopside, cancrinite.

Millionths of a mm Retardation	Interference of Colors Between X Nicols.	$n_g - n_p$	Rock-forming Minerals.
1079	Dark orange violet.	0.036	Olivine, lazurite.
1228	Violet (Sensitive tint No. 2).	0.037	. . . .
1140	Indigo.	0.038	Epidote.
1260	Greenish blue.	0.042	Muscovite.
1300	Sea green.	0.044	Phlogopite, anhydrite.
1425	Greenish yellow.	0.048	Limonite.
1495	Rose red.	0.050	Talc.
1652	Violet gray (Sensitive tint No. 3).	0.056	. . . .
1845	Greenish gray.	0.062	Zircon.
2170	Colors very faint.	0.072	Hornblende.
2900	. . . .	0.097	Cassiterite.
3600	. . . .	0.121	Titanite.
4600	. . . .	0.155	Aragonite.
5200	Not distinguishable.	0.172	Calcite.
5400	. . . .	0.179	Dolomite.
6100	. . . .	0.202	Magnesite.
7200	. . . .	0 239	Siderite.
8400	. . . .	0.280	Hematite.
8600	With shades of red and green.	0.287	Rutile.

The lowest colors of the above scale are the colors of the first order, which includes all of the colors up to the first violet, which marks the limit of the order. The colors of the second and third orders are successively higher. In the fourth order, the colors begin to approach white light, due to an overlapping of the interference. The highest color which the mineral is capable of producing is usually taken for comparison with the colors of the table. In uniaxial minerals, such colors are given by sections cut parallel to the optic axis.

Determination of the Order of Color Produced by Interference.—The rank of an interference color may be determined by means of a "quartz wedge." This is a quartz plate of varying thickness, which gives the colors of the Newton scale from the grayish blue of the first order up. The quartz wedge is mounted on a plate of gypsum in such a position that the faster ray in the gypsum is the slower ray in the quartz. The gypsum plate is made of such thickness that its effect is completely compensated by that of the wedge at the middle of the latter. Between crossed nicols, darkness will result at this point. Upon moving the wedge in either direction, the colors rise successively from this zero or compensating point to colors of the third order. When the wedge is superimposed over the thin section of a mineral, the colors rise in the scale if moved in one direction and fall if moved in the other direction.

Assume that a mineral is placed upon the stage of a microscope between crossed nicols and in its position of maximum illumination. Insert a quartz wedge in the proper slit in the microscope tube above the thin section, and note the change of colors as the wedge is moved over the field, the thin edge being inserted first. If the colors rise in the scale from yellow to red to violet to blue to green and again to yellow, it is an indication that the greater axis of ether vibration of the thin section and that of the quartz wedge are parallel. Therefore, turn the stage 90 degrees to its former position and insert the wedge again. The order of the change of colors will now be reversed. The colors will fall, indicating that the lesser axis of ether vibration of the thin section is parallel to the greater axis of the wedge. Move the wedge over the mineral until the plate becomes dark or gray. This is the "compensation point," where the acceleration of one of the rays of the plate corresponds exactly to the retardation of the same in the wedge. Remove the mineral from the stage. The interference color that the wedge displays is now the same as that originally shown by the mineral. Slowly remove the wedge, observing carefully the sequence of colors. The number of times that any color recurs until the wedge is removed gives the order of the original interference color of the mineral.

From a birefringence chart it is possible to determine not only the order of birefringence of a mineral but the thickness of the section, provided some mineral contained in the slide is known. Let us take, as an example, granite in which quartz is easily recognized. It is fairly safe to assume that, if there are many fragments of quartz in the field of view, the fragment with the highest interference color is cut parallel to the optic axis, and its birefringence has a maximum value, 0.009. This value on the color chart is marked by a diagonal line, which should be followed toward the lower left-hand corner to the intersection with the vertical, giving the interference color shown in the slide. The ordinate at the point of intersection represents the thickness of the section. Its value is determined by following the horizontal line through the intersection to the scale on the left. This reading gives the thickness of the section in millimeters.

Having thus determined the thickness of the section, find again the highest interference color in a fragment of the mineral which is to be determined. Take the intersection of the horizontal line of thickness in the chart with this color. The diagonal line passing through this point of intersection indicates the birefringence of the mineral in question.

Double Refraction of Biaxial Minerals.—Minerals of the orthorhombic, monoclinic and triclinic systems.

Minerals of these three systems have two optic axes,

or two directions in which light is singly refracting; hence the name biaxial. Sections cut perpendicular to these directions remain dark between crossed nicols during a complete revolution of the stage. The optic axes of biaxial minerals never coincide in position with any of the crystallographic axes as is true in the case of uniaxial minerals. In the orthorhombic system they lie in the same plane with two of these axes.

Biaxial minerals contain greatest and least axes of ether elasticity, which, as in uniaxial minerals, are denoted by X and Z. In addition, there is a mean axis of ether elasticity, denoted by Y.

OPTIC AXIAL PLANE. The plane containing X and Z also contains the two optic axes, and is called the optic axial plane, or the optic plane. The mean axis of ether elasticity, Y, is normal to this plane, and is called the optic normal.

BISECTRICES. The optic axes intersect each other at the point of intersection of the optic plane with the other planes of symmetry, if any exist, making equal angles on opposite sides of the axes X and Z. Therefore, X and Z are known as bisectrices. When X bisects the acute angle of the optic axes, it is called the acute bisectrix. The same is true for Z. When they bisect the obtuse angle, they are called obtuse bisectrices.

Orthorhombic minerals have three axes of ether vibration parallel to the crystallographic axes. The direction of X may be the same as a, b or c, the directions of Y and Z varying accordingly, but always at right angles to each other.

In monoclinic minerals, one of the axes of ether vibration, frequently Y, coincides with the crystallographic axis b (the axis of symmetry), and the other two are in the plane of symmetry, parallel to the clinopinacoid.

In the triclinic system, the axes of ether vibration

have no fixed relation to the crystallographic axes.

The discussion of the determination of birefringence of uniaxial minerals is applicable to biaxial minerals. The interference colors in sections of biaxial minerals normal to the optic elements grade downward in the following order from highest to lowest:

1. Optic normal.
2. Obtuse bisectrix.
3. Acute bisectrix.
4. Optic axis.

The following scale of birefringence (after Winchell) is useful for comparison in the estimation of the birefringence of an unknown mineral.

SCALE OF BIREFRINGENCE.

1. Very weak birefringence   0.0035 or less.   Example, leucite.
2. Weak birefringence   0.0035–0.0095.   Example, orthoclase.
3. Moderate birefringence   0.0095–0.0185.   Example, hypersthene.
4. Rather strong birefringence   0.0185–0.0275.   Example, augite.
5. Strong birefringence   0.0275–0.0355.   Example, diopside.
6. Strong birefringence   0.0355–0.0445.   Example, muscovite.
7. Very strong birefringence   0.0445–0.0565.   Example, aegirite.
8. Extreme birefringence   0.0565.   Example, titanite.

CHAPTER 4.

General Methods of Mineral Determination (Continued).

General Operation No. 4: Determination of the Axial Interference Figures.

Interference figures are obtained by the use of crossed nicols in convergent light. A high-power objective must be used. When the eyepiece is removed, a small image of the interference figure can be seen. By sliding the Bertrand lens into the tube of the microscope, a magnified image of the figure is obtained, in which case the ocular is retained. Strong illumination is necessary, with the condensing lens close under the thin section. Results are best with monochromatic light, but the effects are the same with white light except that the rings will be variously colored instead of light and dark.

This operation aids the observer in distinguishing between isotropic, uniaxial and biaxial substances, and aids in the determination of the relative double refraction of minerals.

Isotropic minerals show no interference figures.

UNIAXIAL INTERFERENCE FIGURES:

a. Sections cut perpendicular to the optic axis or vertical crystallographic axis show a dark cross with or without colored rings. The arms of the cross are parallel to the vibration planes of the nicols, and the figure does not move with the rotation of the section.

b. Sections cut oblique to this position show figures which move about the center of the field. The center of the figure may even be outside of the field, but upon rotation its dark bars may be seen to move across the field. These dark bars remain straight and parallel to themselves.

Fig. 8. Uniaxial figure.

If the obliquity of the section is too great, the bars will show a curvature upon entering the field and upon leaving, but they are straight upon crossing the center of the field. The curvature shifts upon crossing the center from one side to the other, thereby differing from the biaxial figures, in which the bars remain curved in the same direction.

c. If the section becomes so oblique to the optic axis as to approach parallelism to it, the black cross appears to break up into hyperbolas which are symmetrically placed with respect to the optic axes, and then unite to form a dark cross again upon completing the rotation of the section.

Sections which are thick and have a strong double refraction show the cross and rings clearly and sharply outlined, many rings being crowded closely together. Thin sections with weak double refraction show broad crosses and no rings. The observer may thus deduce inferences both in regard to the thickness of the section and the strength of the double refraction.

BIAXIAL INTERFERENCE FIGURES:

a. Sections cut normal to an optic axis show a series of concentric colored curves crossed by a single dark bar. The bar changes into a hyperbola and back into a

bar. Sometimes the curves are not observable. The bar
when straight shows the direction of the optic plane with
which it is parallel.

Fig. 9. Positions of the uniaxial interference figure.

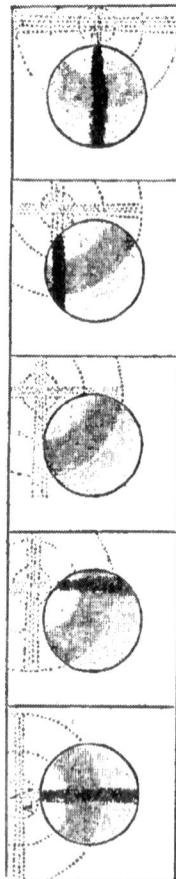

Fig. 9a. Uniaxial dark bars. (Winchell.)

b. Sections cut normal to the acute bisectrix in which
the angle between the optic axes is not too great will
show both optic axes in the interference figure, the bisectrix being in the center of the field between them. In

one case a dark bar appears in the center of the field, its arms varying in size. That line which passes through the optic axes is narrower than the one passing between

Fig. 10.   Optic axis
interference figure.

them. Its extremities widen out on the edge of the field. The intersection of the two bars marks the bisectrix. The trace of the optic plane is the line passing through the loci of the optic axes and the bisectrix. On rotating the section, the dark bars separate into two hyperbolas, the summits receding from each other toward the edge of the field, and beyond it if the optic axes are not in view. They bend through the colored curves surrounding the optic axes, and unite again as a straight bar when

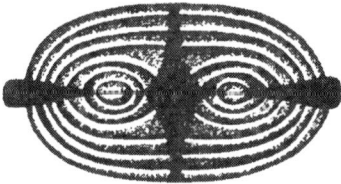

Fig. 11.   Bisectrix interference
figure.

Fig. 12.   Bisectrix interference
figure at 45°.

the plane of the optic axes coincides with the vibration plane of the nicol. The most distant positions of the hyperbolic summits are, therefore, after a revolution of 45 degrees.

An excellent illustration of the biaxial interference figure may be obtained very simply by placing between crossed nicols in convergent light a thin sheet of musco-

Fig. 13. Dark bar with a single optic axis in the field.

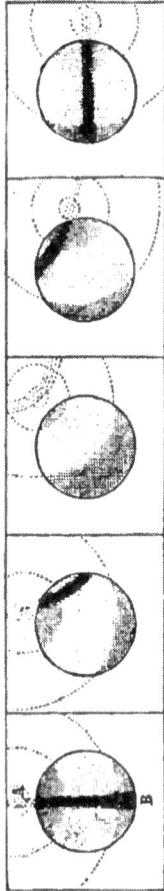

Fig. 14. Dark bar with a single optic axis outside the field.

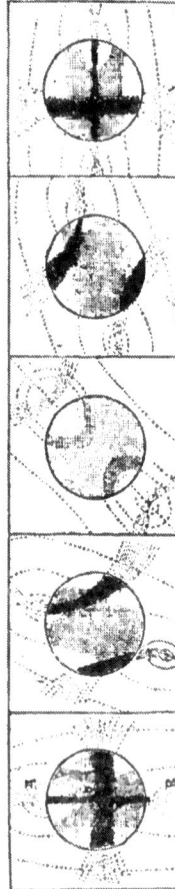

Fig. 15. Changes of the interference figure with a bisectrix in the field.

vite mica. Since the optic angle is small, the loci of both optic axes will be seen in the field. Since the center of the small ellipses and the black hyperbolas mark the loci of the optic axes, they indicate approximately the optic angle.

The uniaxial or biaxial character of a mineral section which shows only an indistinct bar may be determined as follows (La Croix):   .

A bar of a uniaxial interference figure moves in the same direction as the rotating stage, and always remains straight, while the biaxial bar moves in the opposite direction to that of the stage, and becomes curved.

General Operation No. 5: Determination of the Dispersion of the Optic Axes.

The colors of the interference figures in convergent light are caused by the difference of phase of different rays brought together by the analyzer so as to interfere. The phenomenon of relative position of the red and violet rays is caused by the dispersion of the optic axes. When the red ray has the greater optic angle it is expressed by $R > V$; when the violet ray has the greater optic angle, it is expressed by $R < V$.

When white light is used, the colors on the convex side of the hyperbola (which is the side toward the acute bisectrix) are edged with red if the dispersion of red is greater than that of violet, and edged with violet if the reverse is true.

Labradorite, muscovite, orthoclase, and anorthite have a dispersion formula $R > V$. Albite and oligoclase have a dispersion formula $V > R$.

General Operation No. 6: Determination of the Optical Character or the Character of the Double Refraction (after Winchell):

OPTICAL CHARACTER OF UNIAXIAL MINERALS. For light traveling perpendicular to the optic or vertical crystallographic axis, the vibrations of the ordinary ray are transverse to that axis and those of the extraordinary ray are parallel to it; o may be greater or less than e, as the vertical axis may be greater or less than the horizontal crystallographic axes.

If the vertical axis is the direction of the greater axis of ether vibration (X), the mineral is optically negative. The extraordinary ray is less refracted than the ordinary ray, and advances with greater velocity. This is expressed as $o > e$. Optically positive minerals are those in which $o < e$. The greater the velocity, the less the refraction, and the smaller the index of refraction.

To determine the sign of an unknown mineral, one must be able to compare the relative velocities of the ordinary ray (vibrating perpendicular to the primary

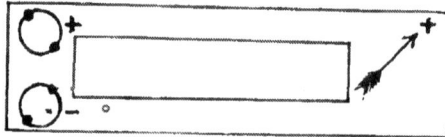

Fig. 16.   Quarter-undulation mica plate.

axis) and the extraordinary ray (vibrating parallel to the primary axis) within the thin section, with the known velocities of another mineral. This may be accomplished by the following methods:

A. With the quarter-undulation mica plate in parallel polarized light.

The quarter-undulation mica plate consists of a cleavage leaf of muscovite, the thickness of which is just sufficient to produce a retardation of a quarter of a wavelength of light, or about two of the larger vertical divisions of the color chart. It is mounted between two glass plates. An arrow on the glass usually indicates the direction of the lesser axis of ether vibration (Z).

When the coinciding axes of the mica plate and the

mineral are the same (Z), the double refraction is increased in proportion to the resultant thickness of the two plates. The color of the section rises through two of the vertical divisions of the color chart. The color falls correspondingly if the axes are not the same.

B. With the quartz-sensitive tint in polarized light.

The quartz-sensitive tint is a plate of quartz cut parallel to its vertical or lesser axis of ether vibration, and is of such a thickness as to give the first violet color of Newton's scale. It is mounted between glass plates. The direction of the vertical crystallographic axis, as well

In parallel polarized light.

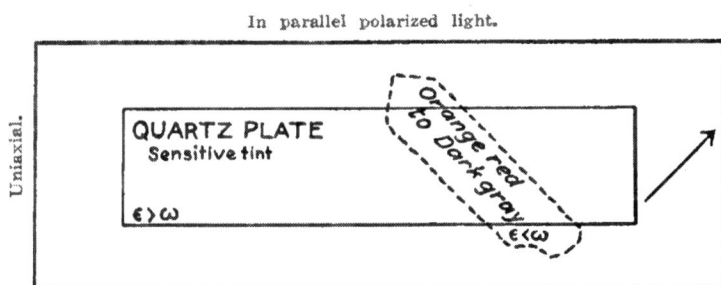

Color falls: Negative.
Fig. 17. Use of the quartz-sensitive tint.

as the optic axis, corresponds with the direction of Z, and is usually indicated by an arrow on the glass.

The position of the axes of ether vibration in the unknown mineral is first determined. This is done by determining on rotation between crossed nicols, the positions at which extinction takes place. The section is placed 45 degrees to this position. The brightest interference color is thus produced. Place the quartz-sensitive tint over the section in such a position that the direction of Z is 45 degrees with the principal sections of the nicols. If by this superposition a color is produced which is higher in the scale than the sensitive tint of the quartz plate, the axis Z of the quartz plate is parallel to the axis Z of the thin section. If the resultant color is lower, the axis of ether vibration of the mineral is X, in consequence of a lessening of the retardation.

Uniaxial minerals are positive when Z coincides with the optic axis, and negative when X does. Since the optic axis coincides with the vertical crystallographic axis, it is necessary when using this method to be able by the crystal outline to determine the direction of the vertical axis. This method is, therefore, practicable only when the section is approximately parallel with the vertical axis and the crystal outline is distinct.

C. With the quartz wedge in parallel polarized light. This method is the same as method B.

Fig. 18.   Disturbance of the interference figure of a uniaxial crystal by the quarter-undulation mica plate.

D. With the quarter-undulation mica plate in convergent light.

By inserting the plate with its Z axis 45 degrees with the cross hairs, the dark cross of the interference figure is destroyed and two dark spots are brought prominently into view. If rings are seen, they will appear disjointed at the lines dividing the quadrants, and they will appear expanded in those quadrants occupied by the dark spots.

The mineral is optically positive if a line joining the two dark spots is perpendicular to the axis of the mica plate. The mineral is negative if the line uniting the dark spots is parallel with the direction of the arrow on the mica plate.

The positive and negative character of the mineral becomes a simple operation if it is borne in mind that the line joining the dark spots makes a positive sign and a negative sign respectively with the axis of the mica plate, thereby indicating directly the sign of the mineral.

E. With the quartz-sensitive tint in convergent light.

Upon inserting the sensitive tint plate, two opposite quadrants will appear yellow and the other set will appear blue. In determining the sign of the mineral, the yellow quadrants may be considered equivalent to the dark spots.

When a section is cut parallel with the optic axis, the interference figure is not a black cross but may resemble a biaxial interference figure. The observer wishes to determine the direction of the optic axis. He may determine this by observing in which quadrants the hyperbolas always leave the field. These will be the quadrants containing the optic axis. Moreover, the interference colors in these quadrants are lower than for corresponding points in the other quadrants. After once determining the direction of the optic axis, Z can be determined by any one of the first three methods for parallel polarized light.

OPTICAL CHARACTER OF BIAXIAL MINERALS. If the greatest axis of ether vibration bisects the acute bisectrix, the mineral is negative. If Z bisects the acute bisectrix, the mineral is positive. Therefore, a determination of the optic sign of a biaxial mineral demands a distinction of the acute from the obtuse bisectrix and a distinction of X and Z.

Distinction between the Acute and Obtuse Bisectrices. The thin section is cut perpendicular to the acute bisectrix if the optic angle is so small that the loci of the optic axes or of one optic axis and the bisectrix remain in the field during a rotation of the stage. Otherwise it is necessary to find sections cut perpendicular to both X and Z and compare them.

1. The section perpendicular to the acute bisectrix shows a lower interference color than the section cut perpendicular to the obtuse bisectrix.

2. The angle of rotation between the position of the black cross and the position when the summits of the hyperbolas are tangent to the edge of the field can be measured. This angle is larger in a section perpendicular to an acute bisectrix than in one perpendicular to the obtuse bisectrix. If the angle is more than 30 or 35 degrees, it is safe to assume that the section is perpendicular to an acute bisectrix. If the angle is less than 15 or 20 degrees, it is perpendicular to an obtuse bisectrix.

Distinction between X and Z. This involves a comparison of the velocities of the light ray in the direction of the axis to be determined with that in the direction of a known velocity in another mineral. Relative retardation is indicated by the relative positions of the colors on Newton's scale. The following methods are available:

A. With the quartz-sensitive tint in parallel polarized light.

The section examined must be parallel to the optic plane, that is, it must contain the axes Z and X. Z of the quartz plate lies in the direction of the arrow, and X at right angles to this direction. Superpose the quartz plate over the mineral. If the resultant color is higher than the sensitive tint of the quartz plate, the Z axes of the quartz plate and of the mineral are coincident. If the resultant color is lower, the Z axis of the quartz plate is coincident with the X axis of the mineral. This enables the observer to determine the position on the two axes.

It is now necessary to view the interference figure in convergent light in order to determine in which quadrants the optic plane lies. If Z is found to lie in the acute optic angle, it bisects the acute bisectrix, and the mineral is positive.

If only two hyperbolas are observed, they are in the quadrants containing the acute bisectrix. If the optic angle is large, hyperbolas may be visible in all four quadrants, but the hyperbolas leave the field more slowly in the quadrants containing the acute bisectrix.

The quarter-undulation mica plate may be used in the same manner as with uniaxial minerals to determine the directions of X and Z since the section is cut parallel to the plane containing X and Z.

B. With the quartz wedge in convergent light.

Obtain an interference figure from a section as nearly normal to the acute bisectrix as possible, and rotate the stage until the optic plane makes a 45-degree angle with the vibration planes of the nicols.

Insert the quartz wedge, with the thin edge advanced, in such a position that the Z axis coincides in direction with a line passing through the optic axes of the figure.

The optical character of the mineral is positive when the ellipses surrounding the loci of the optic axes appear to widen out, and move from the loci of the optic axes toward the center of the interference figure and finally open into the outer colored margins surrounding the whole figure. The optical character of the mineral is negative if the movement of the colors is reversed from the center of the figure toward the axial spots.

With certain interference figures the following simple rule will apply: If the dark spots approach each other, the mineral is negative. If they appear to retreat from each other, the mineral is positive.

C. With the quarter-undulation mica plate in convergent light.

This plate is perpendicular to the negative bisectrix X, and contains Z and Y. The direction of Z coincides with the trace of the plane of the optic axes, since the axial plane always contains X and Z.

For sections perpendicular to an acute bisectrix. When the mica plate is superposed in the usual way, there is an apparent lengthening of the figure in the direction of the Z axis of the mica plate and an apparent shortening in this direction for positive minerals. Winchell suggests that this observation be made with the optic plane parallel with one nicol. The dark spots will now appear in the quadrants through which the arrow passes, the line connecting them forming an angle less than 45 degrees with the arrow, or an approximate minus sign indicating a negative mineral. The reverse takes place with a positive mineral. There is a shortening in the direction of the arrow, and the line connecting the dark spots forms an angle greater than 45 degrees with the arrow, making an approximate plus sign.

Iddings suggests the following method for sections perpendicular to an optic axis: Place the section with its optic plane 45 degrees with the nicols. The hyperbola is convex toward the acute bisectrix. Insert the mica plate with the Z axis parallel with the optic plane of the mineral. The hyperbola moves toward the obtuse bisectrix when the mineral is negative, and toward the acute bisectrix when the mineral is positive. For minerals of weak birefringence, as the feldspars, this method is excellent.

For minerals of strong birefringence the following rule may be applied: The mineral is positive when the black dot appears on the convex side of the hyperbola upon insertion of the mica plate with its Z axis parallel with the optic plane of the mineral.

SUMMARY OF THE OPTICAL SIGN — FOR UNIAXIAL MINERALS. When the E ray is less refracted than the O ray and advances with greater velocity, the mineral is negative, as in calcite. In this case, X coincides with the optic or vertical axis. The index of refraction for the E ray vibrating in this direction is the lesser one, n_p.

When the O ray is less refracted than the E ray and advances with the greater velocity, the mineral is positive, as in quartz. In this case, Z coincides with the vertical or optic axis. The index of refraction of the E ray which is vibrating in this direction is the greater one, n_g.

FOR BIAXIAL MINERALS. When X is the acute bisectrix, the mineral is negative, as in muscovite.

When Z is the acute bisectrix, the mineral is positive, as in augite.

General Operation No. 7: Determination of the Extinction Angle, or the Relation of the Crystallographic Axes to the Axes of Ether Vibration.

This operation is performed between crossed nicols with parallel polarized light.

It will be remembered that the intensity of color depends upon the angle between the principal optic section of the mineral and the principal section of either nicol, the color being greatest at 45 degrees and least at 0 degrees. Thus when the section is in such a position that its directions of elasticity are parallel to the vibration planes of the nicols, no light can pass through the analyzer, and the section is dark. This phenomenon is called extinction.

Extinction is the most common phenomenon for distinguishing isotropic minerals from anisotropic. In the isometric system all minerals are completely dark during a rotation of the stage. It is likewise of great importance in distinguishing between minerals of the three biaxial systems.

Extinction is said to be *parallel* when the directions of the axes of ether elasticity are parallel to any crystallographic directions, which may be determined by cleavage, crystal boundaries, or twinning lines.

Parallel extinction is shown by all sections of tetragonal, hexagonal, orthorhombic minerals, and in the monoclinic minerals in sections parallel to the *b* axis or orthozone. Oblique extinction is shown in all other sections of the monoclinic minerals and all sections of triclinic minerals. In the triclinic system there is no coincidence between the axes of elasticity and the crystallographic axes.

The angle between an axis of elasticity in the section and some known crystallographic direction is called the extinction angle. It is measured as follows: Find the positions of the axes of elasticity in the section when extinction takes place. Note the reading on the vernier of the stage. Rotate the vernier until the known crystallographic direction is brought into a parallel position with the same cross hair which was previously used. A more distinct view of the field may be obtained by removing the upper nicol. The difference between these two readings is the extinction angle.

In monoclinic minerals the maximum value of the extinction angle, which is the only angle of real value in differentiating the mineral, is obtained from a section parallel to the clinopinacoid. Results accurate enough for all practical purposes may be obtained by measuring the angle of all sections of the mineral and taking the maximum value obtained.

Amphiboles and pyroxenes are easily distinguished in this manner.

Extinction which passes over the section like a dark

wave or shadow is called *undulatory extinction.* It indicates that the mineral has been subjected to mechanical forces, producing a change in the position of the axes of elasticity in different parts of the mineral.

It is difficult for the eye to distinguish small variations in the intensity of light. By the use of the quartz-sensitive tint, extinction is determined quite accurately by a distinction of difference of color to which the eye is more susceptible. A thin quartz plate is cut parallel to the axis of elasticity, having such a thickness that it shows the violet color of Newton's scale. Insert this plate in such a position that its axis is 45 degrees to the cross hairs. The field of the microscope is violet. By placing the unknown mineral on the stage so as not to occupy the entire field, it will be seen that the color of the mineral is not the same as the violet color of the unoccupied field. Rotate the stage until the color of the mineral is the violet color of the quartz plate. The mineral is now at extinction. This phenomenon is due to the fact that the axes of elasticity of the nicols and of the mineral are in the same position and producing no interference.

General Operation No. 8: Determination of the Presence or Absence of Pleochroism.

Pleochroism is a property possessed by all anisotropic minerals of absorbing certain colored rays in certain crystallographic directions, thereby showing different colors in different directions by transmitted light. It is observed by polarized or parallel transmitted light.

The axes of absorption coincide generally with the axes of elasticity, therefore with the crystallographic axes in the tetragonal, hexagonal, orthorhombic, and the b axis of the monoclinic systems.

Sections perpendicular to the optic axis can not show

differences in color since in this direction the absorption must be equal in all directions.

Uniaxial minerals are said to be dichroic, showing two different colors, produced by the rays which vibrate parallel to the direction of the vertical axis and parallel to the plane of the basal axes.

Biaxial minerals are said to be trichroic, as there are theoretically three differences in color, corresponding to the directions of the three axes of elasticity. Pleochroism exists practically only in colored minerals.

Pleochroism may be tested as follows: If a mineral is pleochroic, a change in color will be observed upon rotating the stage. This may appear as an actual change in color or as a change in shade of the same color. In case it is almost indistinguishable, it is best to make the test with the condensing lens in position immediately under the section.

An absorption formula is an expression of these different amounts of absorption in any mineral. Thus $a > b$ indicates that absorption is greater when the ether vibration of the polarized ray is parallel to the crystallographic axis a than when parallel to b.

A pleochroic formula expresses the colors that a mineral presents in polarized light vibrating parallel to each of its axes of ether vibration.

For magnesium tourmaline the pleochroic formula is

$Z =$ Dark yellowish brown.

$X =$ Pale yellow.

In general, amphiboles show pleochroism, and pyroxenes do not.

CHAPTER 5.

Description of Important Rock-making Minerals.

INTRODUCTION.

The present-day classification of minerals is primarily a chemical one, as minerals are arranged according to the acid radical. In any chemical division, however, minerals of similar chemical composition, if related crystallographically, are placed in the same group, as for instance the six members of the calcite group.

About a thousand kinds of minerals are known, of which most are rare or found only in a few localities.

Rogers has compiled the following information, which is of interest in a discussion of the derivation of mineral names:

The following minerals were named in honor of prominent scientists: biotite (Biot, French physicist), brucite (Bruce, an early American mineralogist), dolomite (Dolomieu, French geologist), goethite (Goethe, the German poet), millerite (Miller, English crystallographer), scheelite (Scheele, Swedish chemist), smithsonite (Smithson, founder of the Smithsonian Institution), wollastonite (Wollaston, English chemist).

The following minerals were named from prominent geographical localities: andalusite (Andalusia, a province of Spain), aragonite (Aragon, ancient kingdom in Spain), anglesite (Anglesea, in Wales), bauxite (Beaux, in France), ilmenite (Ilmen mountains, in the Urals), labradorite (Labrador), muscovite (Moscow, in Russia), strontianite (Strontian, in Scotland).

The following minerals were derived from the Latin and Greek names for colors: albite (white), azurite (blue), cyanite (blue), celestite (sky-blue), chlorite (green), erythrite (red), hematite (blood), leucite (white), rhodonite (rose-red), rutile (reddish).

The following minerals were named directly from their chemical composition: argentite, arsenopyrite, barite, calcite, chromite, cobaltite, cuprite, magnesite, molybdenite, sodalite, stannite, zincite.

At one time there existed a binomial nomenclature for minerals, as exist at the present time for animals and plants. Thus, barite was known as Baralus ponderosus, and celestite was known as Baralus prismaticus.

The following minerals are discussed according to the crystallographic system in which they occur, isotropic minerals being considered first.

ISOTROPIC MINERALS.

AMORPHOUS.

OPAL.

.. *Composition*: SiO_2. nH_2O.

Criteria for determination in thin section:

Form: No crystal form, but sometimes concretionary, banded or with spherulitic structure.

Optical Properties: $n = 1.45$. Relief so low that the mineral may be mistaken for a hole in the section filled with balsam. Feeble negative double refraction at times. Colorless patches or veins. Fragments are dark and irregular between crossed nicols.

Occurrence: As a secondary mineral in cavities and seams in igneous rocks; as sinter around hot springs and geysers (Yellowstone Park); as a constituent of diatomaceous earth. Diatoms and radiolaria secrete casts of opal silica.

Uses: As gems. Precious opals are found in New South Wales and in Hungary. Fire opal is found in Mexico.

ISOMETRIC

PYRITE.

Composition: FeS_2.

Criteria for determination in thin section:

Form: Cubes, pentagonal dodecahedrons or combinations of these. Sometimes in irregular grains.

Optical Properties: Opaque. In reflected light, pale brass-yellow color with strong metallic luster.

Alteration: Alters readily to limonite by oxidation and hydration.

Occurrence. As a vein mineral associated with other sulphides. As an original and secondary mineral in igneous and sedimentary rocks.

Uses: Used in the manufacture of sulphuric acid. In association with chalcopyrite as a low grade copper ore. It is often gold-bearing.

PYRRHOTITE.

Composition: Fe_6S_7 to $Fe_{11}S_{12}$.

Criteria for determination in thin section:

Form: Practically always in irregular masses and not in crystals. Cleavage usually not visible microscopically.

Optical Properties. Opaque. Color between bronze yellow and copper red. Luster metallic.

Distinctions: Distinguished from pyrite by its usual association in irregular masses and by its bronze yellow color in incident light.

Occurrence: In basic igneous rocks; as a vein mineral; in crystalline limestones.

Uses: The nickeliferous pyrrhotite of Sudbury, Ontario, is an important ore of nickel.

MAGNETITE.

Composition: Fe_3O_4.

Criteria for determination in thin section:

Form: Octahedrons and dodecahedrons. Also granular. Cleavage indistinct. Twinning common after 0.

Optical Properties: Opaque. Bluish black by reflected light, with a strong metallic luster. Index of refraction high.

Alteration: Alters to hematite, limonite and siderite.

Occurrence: A common and widely distributed accessory mineral of igneous rocks; magmatic segregation in ore deposits, as in Scandinavia; as a contact mineral between limestones and igneous rocks; in lenses, in gneisses and schists.

Uses: Important ore of iron, especially in New York, New Jersey and Pennsylvania.

SPINEL.

Composition: $MgAl_2O_4$.

Criteria for determination in thin section:

Form: In grains or octahedral crystals, never decomposed in rocks.

Optical Properties: Strictly isotropic. Index of refraction high. Color usually the lighter shades of red, blue-green, yellow, brown. The most common colors are green (in pleonaste, iron-bearing) and coffee-brown (in picotite, chrome-bearing).

Differentiation: Distinguished from garnets by its octahedral form, by the more common green color, by its undecomposed condition and by its slightly lower relief.

Occurrence: As a contact mineral, in crystalline

limestones and schists. As an accessory mineral, in igneous rocks. In the gem-bearing gravels of Ceylon.

Uses: Ruby-spinel is used as a gem.

GARNET.

Composition: $R''_3R'''_2 (SiO_4)_3$ R'' is Ca, Mg, Fe or Mn. R''' is Al, or Fe.

Criteria for determination in thin sections:

Form: Irregular grains or in simple crystals as dodecahedrons. Zonal structure frequent. No cleavage. Irregular fracture.

Optical Properties: Normally isotropic, sometimes showing anomalous double refraction, due possibly to internal strain.

Colorless or nearly so, to yellowish or reddish. Index of refraction: $n = 1.746$ to 1.814.

Relief high and surface rough.

Alteration: Usually fresh. May be found altered to chlorite.

Differentiation: From spinel, see under the latter.

Occurrence: In schists and gneisses, granites, pegmatites, peridotites, nepheline and leucite-bearing lavas, in crystalline limestones developed at the contact, in sands.

Uses: As an abrasive, particularly for finishing woodwork and leather. Also as a semiprecious gem.

LEUCITE.

Composition: $KAlSi_2O_6$.

Criteria for determination in thin section:

Form: Grains, or well defined, embedded crystals very near the trapezohedron or tetragonal trisoctahedron. Cross sections round or eight-sided. Vary greatly in size. Fine striations due to twinning common. No cleavage, though fracture may be noticed.

Optical Properties:

Colorless. Refringence low. Relief absent. Surface smooth. $n = 1.50$.

Birefringence weak but distinct. Colors of first order (0.001).

Inclusions: Symmetrically or zonally arranged, consisting of the older secretions associated with leucite, magnetite, apatite, augite.

Alteration: Alters readily to zeolites, or mixtures of albite and sericite, or orthoclase and sericite.

Occurrence: Rare in the United States, but common in Italy in basic lavas, substituting the feldspars.

Differentiation: Distinguished from all minerals except analcite and sodalite by its low refringence, crystal form and twinning, very weak birefringence. From leucite and sodalite, by its higher refringence.

SODALITE.

Composition: 3 $NaAlSiO_4$. $NaCl$.

Criteria for determination in thin section:

Form: In concentric nodules. Usually in disseminated or massive form without crystal faces. Crystals if present are dodecahedrons. Fracture conchoidal. Cleavage dodecahedral, generally invisible in thin section.

Optical Properties. Isotropic. May be weakly birefringent around inclusions.

Relief absent. Surface smooth. $n = 1.485$.

Color: Colorless, pink, yellow, blue.

Alteration: Common to fibrous mass of zeolites or to aggregates of micaceous minerals, often accompanied by the formation of limonite and calcite.

Occurrence: In eruptive rocks rich in soda, such as nepheline syenites.

Differentiation: From nephelite by its isotropic

character. From other isotropic minerals, by a very low refractive index.

FLUORITE.

Composition: CaF_2.

Criteria for determination in thin section:

Form: Crystals cubical, octahedral and dodecahedral. Cleavage, perfect octahedral, appearing often in section as triangular cracks.

Optical Properties: Isotropic.

$n = 1.434$. On account of the low index of refraction the negative relief is marked.

Abnormal birefringence may show, due to internal tension.

Color is due to inclusions of hydrocarbons. Some crystals appear green by transmitted light and blue by reflected light. Color not uniformly distributed.

Occurrence: As a very common vein mineral together with calcite, barite, sphalerite, and galena. In limestones. Kentucky and Illinois are chief sources.

Uses: As a flux in iron-smelting and foundry work. Also for the manufacture of hydrofluoric acid and enamels.

CHAPTER 6.

Description of Minerals (Continued).

ANISOTROPIC MINERALS. — UNIAXIAL.

TETRAGONAL.

RUTILE.

Composition: TiO_2.

Criteria for determination in thin section:

Form: Embedded grains, acicular inclusions, massive or in crystals, which are sharp, elongated and prismatic, or in net-shaped groups. Twinning lamellæ common in basal sections. Prismatic cleavage not observed. Elongation parallel to *c*.

Optical Properties: Uniaxial and positive.

Refringence very high. Relief high and surface rough. $n = 2.903$ and 2.616.

Birefringence extreme. Interference colors very high, hence may not be noticed when mineral is strongly colored (0.287).

Extinction parallel to prisms.

Color: Red, brownish red to black.

Pleochroism usually not noticeable. X is yellowish, Z is brownish yellow to yellowish green.

Alteration: Quite stable. May alter to ilmenite.

Occurrence: More widely distributed as a microscopic mineral than as one of megascopic size. Occurs in igneous and metamorphic rocks and in veins. As a secondary mineral in clays. Virginia is source.

Uses: As a source of ferro-titanium and as a coloring matter for porcelain.

ZIRCON.

Composition: $ZrSiO_4$.

Criteria for determination in thin section:

Form: Small, short prismatic crystals usually elongated parallel to c. Always crystallized.

Optical Properties: Uniaxial and positive.

Refringence very high and surface rough. $n = 1.983$ and 1.93.

Birefringence very strong (0.053). Interference colors of fourth order, minute crystals showing brilliant colors.

Color: Colorless to pale gray or brown.

Pleochroism usually not noticeable, but when observed little absorption takes place parallel to c.

Extinction: Parallel to c.

Interference figure in basal section shows several rings in addition to dark cross.

Alterations: Rare.

Differentiations: From apatite, by much higher relief and stronger double refraction. From cassiterite, by much weaker double refraction, and by mode of occurrence.

Occurrence: As an accessory mineral of igneous rocks, especially the more acid varieties. In sands and gravels.

WERNERITE (SCAPOLITE GROUP).

Composition: $mCa_4Al_6O_{25}. + nNa_4Al_3Si_9O_{24}Cl$.

Criteria for determination in thin section:

Form: Crystals rough, coarse and large, in cleavable, columnar and massive forms. Cleavage distinct, parallel to square prism. Elongation parallel to a.

Optical Properties: Uniaxial and negative.

Refringence considerable. $n = 1.583$ and 1.543. Relief not marked and about the same as quartz.

Birefringence rather strong (.03 to .018). Interference colors of the second order more brilliant than those of most of the colored minerals.

Interference figures distinctly uniaxial.

Extinction parallel in longitudinal sections.

Colorless.

Alteration: Alters to kaolinite, muscovite, etc.

Differentiation: From feldspars, by absence of twinning.

• From quartz, by cleavage, higher order of interference colors and optical character.   Quartz is positive.

From apatite, by lower index of refraction, cleavage and higher order interference colors.

Occurrence: Found in gneisses, crystalline schists, and limestones.

HEXAGONAL.

HEMATITE.

Composition: Fe_2O_3.

Criteria for determination in thin section:

Form: Irregular scales, minute grains and earthy. No cleavage.

Optical Properties: Uniaxial and negative.

Refringence very high. $n = 3.042$ and 2.797.

Birefringence very strong (0.245).

Opaque.   By reflected light, black with tinge of red and a metallic luster, or red without luster.

Pleochroism absent, or slight.   X is yellowish red and Z is brownish red.

Alteration: Common by hydration to limonite.

Occurrence: Very widely disseminated.   As microscopic inclusions and as a common alteration product in all rocks. As a commercial iron ore from the Lake Superior district.

ILMENITE

Composition: $FeTiO_3$.

Criteria for determination in thin section:

Form: Irregular masses, without crystallographic outline, or rhombohedral crystals.

Optical Properties:

Opaque. Rarely translucent, and dark brown in very thin sections. Sometimes brownish in reflected light, with metallic luster.

Alteration: To leucoxene, which is believed to be a variety of titanite. This alteration often develops along definite rhombohedral directions.

Differentiation: From magnetite, by occurring in irregular masses and by a whitish strongly refracting decomposition product.

Occurrence: A common though sparsely distributed accessory mineral in igneous rocks and as a magmatic segregation in igneous rocks.

CORUNDUM.

Composition: Al_2O_3.

Criteria for determination in thin section:

Form: Prisms, grains or plates. · Rhombohedral cleavage may show.

Optical Properties: Uniaxial and negative.

Colorless or with patches or zones of blue.

Refringence very high and surface rough. $n =$ 1.7676 and 1.7594.

Birefringence weak like quartz (0.082). Interference colors middle of the first order, yellow to blue.

Interference figure of basal section shows indistinct cross.

Pleochroism marked when color is deep. Z is blue; X is green.

X axis coincides with crystallographic *a*.

Occurrence: In crystalline metamorphic rocks, such as marble, gneisses, mica and chlorite schists, in perido-tites, in sands and gravels.

Uses: Ruby, the red transparent variety, is valuable as a gem. Sapphire, which is likewise valued as a gem, is the blue transparent variety. Burma furnished the best rubies and Ceylon the best sapphires. It is also used as an abrasive.

QUARTZ.

Composition: SiO_2,

Criteria for determination in thin section:

Form: Crystals usually prismatic, terminated by rhombohedrons. Allotriomorphic in granitoid rocks, rounded grains in clastic rocks. Rarely in distinct crys-tals in any rocks. May be mutually interpenetrated by an acid feldspar. Cleavage nearly always absent or difficult.

Optical Properties: Uniaxial and positive.

Colorless. By reflected light it may appear cloudy if it contains many inclusions.

Refringence low. No relief and smooth surface. $n = 1.553$ and 1.554.

Birefringence weak with interference colors of white or yellow in the middle of the first order (0.009).

Pleochroism absent.

Extinction takes place, but is not distinctive, due to the absence of cleavage or crystallographic outline.

Interference figure of a basal section shows a dark cross without any rings.

Alteration: Does not alter, so that the fresh appear-ance of the mineral is an important aid in identification.

Inclusions: Minute fluid or gas inclusions common in granitoid rocks. Not so abundant in porphyritic rocks.

Occurrence: One of the most abundant minerals found in nature. It occurs in sedimentary, acid igneous, metamorphic rocks and veins.

Differentiation: From sanadine, by uniaxial and positive character.

From nephelite, by absence of hexagonal outline, stronger double refraction, and fresh undecomposed appearance.

Uses: For ornamental purposes, for optical instruments, for glass-making, for pottery and porcelain, and as an abrasive.

CALCITE.

Composition: $CaCO_3$.

Criteria for determination in thin section:

Form: Grains and aggregates. May be fibrous or oölitic. Never in crystals in rocks. Polysynthetic twinning common, probably due in part to the grinding of the section. Shows in crossed nicols as a series of light and dark bands. Cleavage parallel to R appearing as many cracks.

Optical Properties: Uniaxial and negative.

Colorless when pure, but may appear colored by transmitted light, due to organic pigments.

Refringence low. Relief not marked and surface smooth. $n = 1.658$ and 1.486.

Birefringence very strong, with pale, iridescent, interference colors of the fourth order (0.172).

Extinction parallel to cleavage cracks when they appear.

Pleochroism. No change of color observed, but absorption can be noted if section is not too thin.

Interference figure of basal section shows distinct cross and rings.

Inclusions of fluid frequent.

Differentiation: From other carbonates, difficult except by microchemical tests.

Occurrence: Abundant in sedimentary limestones and as a decomposition product in igneous rocks. Vein mineral often associated with ores as gangue. As travertine and cave deposits.

DOLOMITE.

Composition: $(Ca.Mg)\ CO_3$.

Criteria for determination in thin section:

Form: In rocks chiefly as crystals, usually unit R, with a tendency to curved surfaces. As dense homogeneous aggregates showing tendency toward crystalline boundaries.

Optical Properties: Uniaxial and negative.

Similar to calcite, from which it may be differentiated by slightly higher relief ($n = 1.682$ and 1.503) and by tendency toward crystalline boundaries.

Occurrence: As the essential constituent of dolomitic limestones, as a vein mineral, and as a secondary mineral in cavities in limestone.

Uses: As limestones for building and ornamental purposes. Also for furnace linings.

SIDERITE.

Composition: $FeCO_3$.

Criteria for determination in thin section:

Similar to calcite in form and optical properties.

Absorption often distinct.

Alteration: Changes readily on exposure to limonite and hematite.

Differentiation: From calcite, by common association with limonite.

From dolomite and magnesite, by common polysynthetic twinning. It is the only mineral of the Calcite group with both indices of refraction higher than that of balsam except the rarer smithsonite and rhodocrosite.

Occurrence: In limestone, clay iron-stone, clay slate, gneiss. Also in veins with metallic ores.

APATITE.

Composition: $Ca_5(Cl.F)(PO_4)_3$.

Criteria for determination in thin section.

Form: Minute, slender, hexagonal prisms, with regular hexagonal boundaries. Grains. Clusters of crystals. Elongation parallel to a. Cleavage seldom observed.

Optical Properties: Uniaxial and negative.

Colorless usually in thin section. Sometimes gray, blue or brown, the color being irregularly distributed, due perhaps to microscopic inclusions.

Refringence moderate. Relief more marked than of the associated colorless minerals. $n = 1.638$ and 1.634.

Birefringence: Weak, with interference colors grayish blue or white, of the lower first order (0.004).

Extinction parallel to c axis.

Interference figure shows a cross without rings.

Pleochroism absent for white crystals. Colored varieties weakly pleochroic.

Alterations: Mineral always appears fresh.

Differentiation: From nephelite, by occurring in smaller and longer crystals, and invariably fresher in appearance.

From zircon, see under the latter mineral.

From feldspars, when granular, by higher relief and the uniaxial interference figure.

From quartz, in having a higher relief, weaker birefringence, and a negative sign.

Occurrence: Widely distributed as an accessory constituent of igneous rocks and in crystalline schists. With metamorphic limestones. As a vein mineral in gabbro and in pegmatites.

Uses: As a source of phosphates for fertilizers.

· NEPHELITE.

Composition: $NaAlSiO_4$.

Criteria for determination in thin section:

Form: Crystals thick, six-sided prisms with base prominent. Massive and in embedded grains. Cleavage imperfect, parallel to the prism of the first order and the base, better in partially altered sections.

Optical Properties: Uniaxial and negative.

Colorless in thin section.

Refringence low. Relief absent. $n = 1.546$ and 1.542.

Birefringence very weak (0.005). Interference colors grayish white of the lower first order, a little lower than the feldspar colors.

Extinction parallel to cleavage lines when they appear.

Pleochroism absent.

Interference figure is a broad cross without rings.

Inclusions: Microscopic needles of augite, also fluid and gas generally in zones.

Alteration: Readily to fibrous zeolites with stronger birefringence.

Differentiation: From quartz, by weaker birefringence, better hexagonal outline, and negative sign.

From feldspars, by uniaxial character and absence of twinning.

Occurrence: In nephelite syenites, phonolites, and rare soda-rich rocks. It is never associated with quartz, but often with orthoclase.

TOURMALINE.

Composition: R_6SiO_5? R chiefly Al, K, Fe, Ca, Mn, Mg, Li.

Criteria for determination in thin section:

Form: Columnar crystals, bunched or in radiating aggregates. Irregular cracks may appear, but no cleavage is seen. Cross section shows trigonal outline parallel to base.

Optical Properties. Uniaxial and negative.

Color: Varies, with grayish blue, brown, and black most common. Zonal structure may be shown by differences in color.

Refringence medium. Conspicuous against the colorless rock constituents. Surface rough, $n = 1.636$.

Birefringence quite strong (0.02), with bright interference colors of the upper first or lower second order. Often masked by strong absorption.

Interference figure shows a sharp cross with a few rings.

Extinction parallel to the c axis.

X axis is parallel to the c axis.

Pleochroism distinct even in light-colored varieties, increasing with the depth of the color. The greatest absorption takes place normal to the direction of elongation of the mineral. Formula for Mg tourmaline Z is pale yellow. X is colorless.

Absorption very marked.

Alteration does not take place commonly.

Differentiation: From hornblende, by absence of cleavage, and by the fact that the greatest absorption takes place at right angles to the longitudinal axis.

Occurrence: Widely distributed in crystalline schists and gneisses, in crystalline limestones (New Jersey), in granite pegmatites and veins with copper minerals. It is a common product of contact metamorphism.

Uses: Colored tourmaline is used as a gem.

CHAPTER 7.

Description of Minerals (Continued).

ANISOTROPIC-BIAXIAL MINERALS.

ORTHORHOMBIC.

ANDALUSITE.

Composition: Al_2SiO_5.

Criteria for determination in thin section:

Form: Prismatic crystals always more or less elongated parallel to the vertical axis, in rough or embedded crystals. Cleavage may show parallel to almost square prism.

Optical Properties: Biaxial and negative.

Color: Colorless to reddish.

Refringence medium. $n = 1.64$ and 1.63.

Birefringence weak (0.01). Interference colors, middle of the first order, white or yellow.

Interference figure shows large optic angle.

Extinction parallel to c.

Pleochroism marked only in colored varieties, being reddish parallel to c, which is the direction of elongation or cleavage. Pleochroic halos often surround inclusions.

Inclusions: Carbonaceous matter common, distributed through the crystal in some geometrical form conforming to the symmetry.

Alteration: Readily to colorless mica.

Differentiation: From diopside, by weaker birefringence and absence of extinction angles.

Occurrence: In granitic eruptive rocks and in meta-morphosed sedimentary limestones.

TOPAZ.

Composition: $Al_2F_2SiO_4$.
Criteria for determination in thin section:
Form: Colorless crystals of short prismatic habit. Cleavage perfect parallel to the base.
Optical Properties: Biaxial and positive.

Refringence medium, about the same as calcite. $n = 1.617$ and 1.607.

Birefringence weak (0.01), about the same as that of quartz with interference colors, middle of the first order white and yellow.

Interference figure shows large optic angle.

Extinction parallel to cleavage.

Z axis parallel to *c*.

Alteration: To kaolin or muscovite by loss of F and addition of water and alkalies.

Differentiation: From quartz, by cleavage and biax-ial character.

From andalusite, by its cleavage and its smaller optic angle.

From orthoclase, by its higher relief, absence or rarity of twinning and extinction parallel with the cleavage.

Occurrence: In contact metamorphic zones and in pegmatite, associated with cassiterite, fluorite, tourma-line, beryl, etc.   In cavities in rhyolite.

Uses: Occasionally as a gem.

STAUROLITE.

Composition: $FeAl_5(OH)(SiO_6)_2$.
Criteria for determination in thin section:
Form: Short prisms twinned at 90 or 60 degrees. Cleavage variable, both prismatic and pinacoidal.

Optical Properties: Biaxial and positive.

Color: Yellowish to brown.

Refringence rather high and surface rough. $n = 1.746$ and 1.736.

Birefringence weak (0.01) with interference colors middle of first order white to yellow, about like quartz.

Optic angle large. Plane of optic axis is parallel to 100.

Pleochroism distinct but not strong, showing the darker color parallel to c, the direction of elongation (Z is golden yellow, Y is pale yellow, X is colorless).

Extinction parallel to cleavage or crystal outline.

Inclusions of rutile, tourmaline, garnet and quartz occur, the latter abundantly.

Alteration: To a green mica and chlorite.

Differentiation: From titanite, by the fact that in convergent light the optic plane is shown to be in the longer diagonal of the cross section.

.Occurrence: In mica schists and phyllites associated with garnet, cyanite and andalusite.

SERPENTINE.

Composition: $H_4Mg_3Si_2O_9$.

Criteria for determination in thin section:

Form: Not known in crystal form. Fibrous or scaly masses with elongation parallel to c. Prismatic cleavage of 130 degrees seldom visible.

Optical Properties: Biaxial and positive.

Color in thin section: Pale green, yellow, or colorless.

Refringence low, about the same as Canada balsam. No relief, and smooth surface. $n = 1.54$.

Birefringence rather weak, with interference

colors middle of the first order, gray, white or yellow. Between crossed nicols the aggregate structure is distinctly seen. Fine fibrous aggregates may appear isotropic (0.013).

Pleochroism in thick sections distinct. Z is green or yellow. Y and X are greenish yellow to colorless.

Optic plane parallel to 010. Optic angle is small.

Extinction parallel.

Differentiation: From chlorite, by its weaker pleochroism.

From fibrous amphiboles, by much weaker birefringence, lower relief and parallel extinction. Fibrous structure and color indicative.

Occurrence: As an alteration product of olivine, amphiboles, pyroxenes. The essential mineral in the metamorphic rock serpentine, derived from peridotite. A secondary occurrence in veins.

Uses: As an ornamental stone. The fibrous variety forms a commercial asbestos.

THE ORTHORHOMBIC PYROXENES.
ENSTATITE AND HYPERSTHENE.
ENSTATITE.

Composition: $MgSiO_3$.

Criteria for determination in thin section:

Form: Distinct crystals rare, prismatic. Columnar or fibrous structure parallel to *c*, characteristic of allotriomorphic occurrences. Usually massive, fibrous or lamellar. Prism angle nearly 90 degrees. Twinning not as common as in the monoclinic pyroxenes. Prismatic cleavage distinct.

Optical Properties: Biaxial and positive.

Color: Colorless in thin sections. Bronzite, which is a variety of enstatite containing ferrous iron in place of some of the magnesium, is colorless or nearly so, and shows strong pleochroism with X a pale yellow, Y a brownish yellow, and Z a bright green.

Refringence high and surface rough, about the same as in the monoclinic pyroxenes. $n = 1.665$ and 1.656.

Birefringence weak (0.009) much weaker than the monoclinic pyroxenes. Interference colors low of first order.

Interference figures not marked on account of the weak double refraction.

Extinction parallel to cleavages, both pinacoidal and longitudinal prismatic, and bisecting angles of intersecting prismatic cleavages.

Axial plane parallel to brachypinacoid, that is, parallel to the best cleavage. Axial angles large.

Pleochroism weak or absent.

Alteration: To serpentine by ordinary weathering. Also to uralite (a variety of hornblende), but much less commonly than the monoclinic pyroxenes do.

Differentiation: From hypersthene, by the optic sign and absence of distinct color and pleochroism.

From the monoclinic pyroxenes, by parallel extinction on vertical sections, and lower interference colors.

Occurrence: A common constituent of basic igneous rocks as well as of serpentine derived from them. Also found in crystalline schists and in many meteorites. Bronzite contains about 10 per cent FeO and has a characteristic bronzy luster due to inclusions.

HYPERSTHENE.

Composition: (Mg, Fe) SiO$_3$.

Criteria for determination in thin section:

Form: Similar to enstatite. More often massive in lamellæ. Elongated parallel to *c*.

Optical Properties: Biaxial and negative.

Color: Brownish to greenish.

Refringence slightly higher than enstatite, due to increase in percentage of iron.

Birefringence slightly stronger than enstatite, due to increase of iron. Weaker than monoclinic pyroxenes.

Extinction same as enstatite.

Axial plane parallel to brachypinacoid, i.e., parallel to the best cleavage. Optic angle about X becomes smaller with increase in iron content.

Pleochroism distinct, increasing with increase in iron. Z is bright green, Y is yellowish brown, X is clear red.

Inclusions: Gaseous, liquid, glassy. Also a reddish brown material regularly arranged, which gives it a peculiar submetallic bronze-like luster. They are believed to be inclusions of ilmenite, either primary or produced at depth under pressure by circulating waters acting along a cleavage or parting plane.

Alteration: To a variety of serpentine called bastite, less commonly to uralite, occasionally to talc.

Occurrence: Important constituent with plagioclase in basic igneous rocks, as norites and gabbros. Abundant in andesites. Found in meteorites.

BASTITE.

Bastite is a variety of serpentine to which the orthorhombic pyroxenes poor in iron alter frequently through the ordinary processes of weathering. It is geometrically oriented on the altered pyroxene, replacing crystal for crystal. It is composed of fibers often traversed by irregular cracks. Cleavage traces of the two minerals coincide, but the optical properties differ. The pyroxene has a cleavage parallel to the trace of the optic plane. In bastite, the cleavage is perpendicular to the trace of the optic plane, and to the negative acute bisectrix. This is the surest distinction between them.

Bastite is light yellowish or greenish. Refringence is less than that of the orthorhombic pyroxenes and about the same as Canada balsam. Birefringence is weak. Extinction is parallel to the fibres. Pleochroism is weak and seen only in thick sections.

OLIVINE (CHRYSOLITE).

Composition: $(Mg. Fe)_2 SiO_4$.

Criteria for determination in thin section:

Form: Idiomorphic, or in grains or granular aggregates. Also massive. Longitudinal sections more or less lath-shaped with pointed ends. Outlines of crystals often corroded or rounded. Interpenetration twins occur. Cleavage distinct, parallel to brachypinacoid, less distinct parallel to macropinacoid, often made more visible by decomposition. An irregular fracturing is often conspicuous, especially where alteration to serpentine has commenced. Elongation usually parallel to *c*.

Optical Properties: Biaxial and positive.

Color: Nearly colorless, becoming reddish with high iron content.

Refringence high. Relief marked and surface rough. $n = 1.689$ and 1.6535.

Birefringence very strong, with interference colors of the second or third order, higher than the colors of augite (0.0359).

Extinction always parallel to cleavage cracks.

Axial plane parallel to the base, that is, at right angles to the general direction of elongation. Axial angle very large.

Pleochroism absent except in reddish varieties.

Inclusions: Magnetite, spinel, apatite, common. Also liquid or gas.

Alteration: Alters readily. Altered forms are more frequently observed than the fresh. Serpentine is the commonest alteration product, with frequently a separation of magnetite or hematite. The first alteration goes on along the cleavage and fracture cracks.

It is easily altered by atmospheric weathering to carbonates with limonite and opal or quartz. Calcite may usually be distinguished in this case. In contact with a feldspar it may alter to an amphibole by regional metamorphism. The amphibole appears as a zone of pale green or colorless needles between the olivine and the feldspar.

Differentiation: From light colored monoclinic pyroxenes by parallel extinction, by poorer and unequal cleavages and stronger birefringence. Olivine should be easily recognized by its high refringence, a shagreen surface, no color, strong birefringence, and a large optic angle.

Occurrence: Especially in basic igneous rocks, associated with augite, hypersthene, plagioclase, magnetite. An essential constituent of many meteorites, constituting the stony portion of the mass.

Uses: The transparent variety is sometimes used as a gem under the name peridot.

TALC.

Composition: $H_2Mg_3(SiO_3)_4$.

Criteria for determination in thin section:

Form: Colorless plates, elongated like rods. More rarely with round to hexagonal outline. May be arranged in rosettes. Often more or less compact foliated masses.

Cleavage perfect parallel to the base like mica. Elongation parallel to *c*.

Optical Properties: Biaxial and negative.

Colorless in thin section.

Refringence moderate. $n = 1.589$ and 1.539.

Birefringence strong, with interference colors of the third order, like muscovite (0.05 to 0.035).

Extinction parallel to basal cleavage lines.

Plane of optic axes parallel to 100.   Optic angles small.

Differentiation: From muscovite, by its small optic angle. From sericite, by lower refringence.

Occurrence: Most abundantly in crystalline schists, often forming rock masses, as soapstone. As a secondary mineral in basic igneous rocks, altering from olivine, enstatite, tremolite.

Uses: For soap, talcum powder, French chalk, and in the manufacture of paper.

NATROLITE.

Composition: $Na_2Al_2Si_3O_{10}. 2H_2O$.

Criteria for determination in thin section:

Form: Aggregates of colorless, fibrous crystals, often in interlacing groups or divergent.   Prismatic angle nearly 90 degrees. Elongation parallel to c. Microscopic twinning on 110.

Optical Properties: Biaxial and positive.

Refringence very low, with no relief. $n = 1.485$ and 1.473.

Birefringence weak, though slightly stronger than quartz.

Interference colors, middle of the first order yellow, a little higher than quartz.

Interference figure shows dark cross. Optic angle large.

Plane of optic axis parallel to 010.

Axis Z parallel to c.

Extinction parallel to fibers.

Occurrence: Never found as a primary mineral. As a secondary mineral in basic igneous rocks filling amygdaloidal cavities. Common alteration product of sodalite, nephelite, and acid plagioclase.

PYROXENE GROUP.

Composition: The pyroxenes are metasilicates of calcium, magnesium, iron, or more complex silicates, often containing two or more bases, both bivalent and trivalent. They are closely related to each other in crystallographic and physical properties. They crystallize in the orthorhombic, monoclinic and triclinic systems.

Criteria for determination in thin section:

Form: Fundamental form is a short prism with interfacial angles of about 87 and 93 degrees. Distinct cleavage occurs parallel to both prism faces. Twinning if present is parallel to 100. Elongation usually parallel to *c*.

Optical Properties: Biaxial. Most species positive.

Color: In thin section usually pale to colorless. Soda pyroxenes have a distinct green color.

Refringence high. Relief distinct and surface rough. n = 1.68 to 1.72.

Birefringence strong, being stronger in the pale or colorless pyroxenes with interference colors, bright tints of the second order '(0.021 to 0.030).

Extinction: Maximum angle from 0 to 95 degrees, with common species varying between 30 and 54 degrees. In sections showing parallel cleavage lines, parallel extinction in orthopinacoidal sections, and in all other sections an extinction angle is observed. The maximum extinction angle is large and is obtained only when the section of the crystal is parallel to the clinopinacoid.

Optic plane is parallel to 010.

Pleochroism weak or absent except in the soda pyroxene.

Alteration: Alters readily to amphiboles. Described under each species.

Differentiation: Pyroxenes may be distinguished from amphiboles by the following criteria:

Pyroxenes.	*Amphiboles.*
Cleavage angle about 93 degrees.	Cleavage angle about 124 degrees.
Crystals short prismatic.	Crystals long prismatic.
Color usually weak.	Color marked.
Pleochroism weak.	Pleochroism marked.
Extinction angles 0–95 degrees.	Extinction angles 0–25 degrees.
Most species positive.	Most species negative.
Alter to amphiboles.	Alter to chlorite, biotite, etc.

MONOCLINIC MINERALS.

Monoclinic Pyroxenes.

Diopside.

Diallage.

Augite.

Aegirite.

DIOPSIDE.

Composition: $Ca(Mg, Fe)(SiO_3)_2$.

Criteria for determination in thin section:

Form: Long, columnar crystals and grains. Often coarsely lamellar. Granular masses. Cleavage always distinct parallel to 110 in two directions nearly at right angles to each other. Parting parallel to the base, yielding fine twinning lamellæ in this direction. Elongation parallel to c.

Optical Properties: Biaxial and positive.

Colorless usually in thin section; but with increase in iron, color becomes distinctly greenish.

Refringence increases with increase in iron content. $n = 1.7026$ and 1.6727.

Birefringence decreases with increase in iron content (0.0299).

Interference colors bright, of the second order.

Interference figure is an axial bar with concentric rings.

Extinction angle from 20 to 30 degrees.

Plane of the optic axes parallel to 010.

Dispersion weak.

Inclusion: Gaseous, liquid, or glassy, arranged in zones.

Alteration: Most commonly to serpentine or to an aggregate of serpentine and chlorite, often with calcite and quartz.   Also to actinolite or hornblende, alteration starting around the periphery or along cleavage cracks.

Differentiation: From augite, by less dispersion, consequently better extinction in white light.

From ægirite and spodumene, by the extinction angle in the vertical zone.

From hypersthene, by the absence of pleochroism.

From orthorhombic pyroxenes, by the extinction angle and the higher order of colors.

Occurrence: In crystalline limestones as a contact mineral with garnet.   In many igneous rocks as granites, diorites, syenites, gabbros, and peridotite.   In metamorphic rocks.

DIALLAGE.

A variety of diopside showing well developed parting parallel to 110, generally showing a fibrous texture parallel to c.   Color usually brown, with pleochroism as follows: Z is greenish, Y is brownish or reddish brown, X is greenish. It contains inclusions like those of hypersthene, which gives it a bronze-like luster.

AUGITE.

Composition: $mCaMg (SiO_3)_2 . n(Mg, Fe) (Al,Fe)_2 SiO_6$.

Criteria for determination in thin section:

Form: Crystals short thick prisms coarsely lamellar, parallel to 001 or 100. Granular. Twinning common, giving polysynthetic lamellæ parallel to 100. Cleavage imperfect, but distinct in two directions parallel to 110 nearly at right angles. Elongation parallel to *c*.

Optical Properties: Biaxial and positive.

Color green, greenish black, brown. Rarely yellow.

Refringence high. High relief and rough surface. $n = 1.733$ and 1.712.

Birefringence rather strong, being stronger in the pale or colorless pyroxenes. Interference colors are bright tints of the second order (0.021).

Interference figures distinct on account of the strong birefringence. Axial angles large.

Optic plane parallel to the clinopinacoid (010).

Extinction angle: Maximum from 38 to 51 degrees, which is obtained when the section of the crystal is parallel to the clinopinacoid (010), varying from these angles to 0 degrees when the section is parallel to the orthopinacoid (100).

Pleochroism usually absent or weak unless rich in iron, in which case Z is greenish, Y is brownish to reddish brown, and X is green.

Inclusions: Gaseous, liquid or glassy, sometimes arranged in zones.

Alteration: Most commonly to uralite, a variety of hornblende, either crystal for crystal, or to a fibrous aggregate of uralite.   The alteration begins around the periphery of the crystal or along cleavage cracks.   It may alter to biotite and then to chlorite, or directly to chlorite, sometimes forming calcite, quartz or epidote simultaneously.

Differentiation: From diopside, see under the latter.

From ægirite and spodumene in the extinction angle in the vertical zone.

From amphiboles, see under Pyroxene group.

From epidote, by the fact that the plane of the optic axis is parallel to the longitudinal axis and cleavage cracks.

Occurrence: Abundant in igneous rocks, but found also in metamorphic rocks.   Occurs in some stony meteorites.

ÆGIRITE (ACMITE).

Composition: Na Fe $(SiO_3)_2$.

Criteria for determination in thin section:

Form: In crystal form, similar to augite, although often longer or acicular.   Cleavage parallel to 110, more distinct than· in augite, almost at right angles.   Parting parallel to 100.   Elongation parallel to c.

Optical Properties: Biaxial and negative.

Color in thin section greenish or brownish.

Refringence high, with high relief and rough surface. $n = 1.8126$ and 1.762.

Birefringence quite strong, with bright tints of the second order, although it is stronger in the pale or colorless varieties of pyroxene (0.0496).

Interference figures distinct on account of strong birefringence.   Optic angle is large.

Optic plane parallel to the clinopinacoid (010).

Extinction angle: About 5 degrees.

Pleochroism marked. Z is yellowish green, Y is olive green, X is dark grass green.

Alteration: To analcite and to the iron oxides.

Differentiation: From the amphiboles, see under pyroxene group.

From other monoclinic pyroxenes, by the very small extinction angle, the negative sign, the stronger birefringence and marked pleochroism.

Occurrence: In pegmatite veins, in soda-rich igneous rocks, as nephelite, syenites, phonolites and soda-granites.

AMPHIBOLE GROUP.

Composition: The minerals of the amphibole group are orthorhombic, monoclinic, and triclinic silicates of magnesium, calcium iron or sodium, with aluminum or ferric iron in some cases.

Criteria for determination in thin section:

Form: Crystals usually prismatic, elongated parallel to c, possessing very marked and regular prismatic cleavages, varying little from 124 degrees between the cleavage faces. Twinning common parallel to 100.

Optical Properties: Biaxial. Most species negative.

Color in thin section are green, brown, blue, yellow or colorless.

Refringence averages less than the pyroxenes, increasing with increase of iron. Relief distinct. n = 1.621 to 1.642.

Birefringence quite strong, but a little weaker than in the pyroxenes. Interference colors are bright tints of the second order. May be masked by strong absorption (0.019 to 0.027).

Extinction: Maximum angle from 0 to 25 degrees.

In the monoclinic amphiboles the axis Z makes an angle with the vertical crystallographic axis *c*, which varies from 0 to 22 degrees, except in uncommon species. Elongation therefore positive.

Pleochroism distinct and intense in the colored varieties.

Absorption very marked, being greatest in the general direction of the cleavage lines in the longitudinal sections (parallel to Z).

Optic plane parallel to 010 in monoclinic amphiboles.

Inclusions: Iron ores, apatite, etc.

Alteration: Alter readily to chlorite, biotite, sericite, epidote, calcite, talc, etc., the process being gradual, and usually beginning along the edges and the cleavages of the amphibole until all trace of the original mineral is lost.

Differentiation: Amphiboles may be distinguished from the pyroxenes by the following criteria:

Amphiboles.	*Pyroxenes.*
Cleavage angle about 124 degrees.	Cleavage angle about 93 degrees.
Crystals long prismatic.	Crystals short prismatic.
Color marked.	Color usually weak.
Pleochroism marked.	Pleochroism weak.
Extinction angles, 0–25 degrees.	Extinction angles, 0–95 degrees.
Most species negative.	Most species positive.
Alter to chlorite, biotite, etc.	Alter to amphiboles.

Occurrence: In all classes of eruptive rocks and in many metamorphic rocks. Often formed by alteration from pyroxenes.

Monoclinic Amphiboles:

Tremolite.
Actinolite.
Hornblende.
Riebeckite.

TREMOLITE.

Composition: Ca $Mg_3(SiO_3)_4$.

Criteria for determination in thin section:

Form: Crystals long-bladed or short prismatic. Often fibrous or acicular. Perfect prismatic cleavage at an angle of about 124 degrees. Cleavage sometimes distinct, parallel to 010 and 100. Transverse fracture frequent. Cleavage more perfect than in pyroxenes.

Optical properties: Biaxial and negative.

Colorless in thin section.

Refringence high, with distinct relief, but not as marked as in the pyroxenes. $n = 1.634$ and 1.6065.

Birefringence quite strong, but a little weaker than in pyroxenes (0.0275).

Optic plane parallel to 010.

Maximum extinction angle is 18 to 16 degrees in vertical zone.

Dispersion weak.

Inclusions of carbonaceous matter and biotite in tremolite of metamorphic rocks.

Alteration: To talc, beginning along cleavage lines. Also to calcite.

Differentiation: From hornblende, by light color. It has the lowest index of refraction found in monoclinic amphiboles.

From pyroxenes, see under Amphibole group.

Occurrence: In schists, contact rocks, and veins.

Uses: Fibrous varieties sometimes used as asbestos. As jade, sometimes used for ornamental purposes.

ACTINOLITE.

Composition: $Ca(Mg, Fe)_3 (SiO_3)_4$.

Criteria for determination in thin section:

Form: Similar to tremolite.

Optical Properties: Biaxial and negative.

Color in thin section pale to dark green, depending upon the percentage of iron.

Refringence and birefringence similar to tremolite (0.025).

Maximum extinction angle in vertical zone is 15 degrees.

Dispersion weak.

Pleochroism pronounced, and absorption marked, being greatest in the general direction of the cleavage lines in longitudinal sections. Z is pale to dark green, Y is greenish yellow, X is very pale yellow.

Inclusions: Similar to tremolite.

Alteration: To chlorite, epidote, talc, etc.

Occurrence: Same as tremolite, with which it is often associated. Uralite is the name given to the amphibole to which the pyroxenes frequently alter. It usually corresponds to actinolite.

HORNBLENDE.

Composition: $mCa(Mg, Fe)_6 (SiO_3)_4$. $n(Al, Fe)$ $(F, OH)SiO_3$.

Criteria for determination in thin section:

Form: Prismatic elongated, parallel to the vertical axis, sometimes fibrous. Prismatic cleavage perfect, making the characteristic angle of 124 degrees. Parting and polysynthetic twinning are sometimes present, parallel to 100 or 001. Cross sections may be acutely rhombic, with acute angles truncated, hence six-sided, whereas the pyroxenes are usually eight-sided. Longitudinal sections lath-shaped. Zonal structure occurs fre-

quently in the brown hornblende. Twinning frequently
parallel to the orthopinacoid.

Optical Properties: Biaxial and negative.

Colorless, gray, green, greenish blue, brown or
black.

Refringence high, with distinct relief. $n = 1.653$
and 1.629.

Birefringence quite strong, being strongest in
the basaltic hornblende. In common hornblende,
$n = (0.024)$. In basaltic hornblende, $n = (0.072)$.
Interference colors, bright tints of the second order,
often masked by strong absorption.

Maximum extinction angle in common hornblende
20 degrees, in basaltic hornblende from 1 to 2 degrees.

Optic plane parallel to 010, in which face Z makes
a variable angle with the axis c in the obtuse angle
Beta.

Dispersion distinct.

Pleochroism: Z pale green, Y pale brown, X clear
brown, for common hornblende.

Inclusions abundant but not characteristic. Rutile
common.

Alteration: By ordinary weathering to chlorite often
accompanied by epidote, calcite and quartz. Sometimes
to biotite. By heat to augite.

Differentiation: From other amphiboles, by stronger
color and pleochroism, higher interference colors.

From pyroxenes, see under Amphibole group.

Occurrence: Widespread in igneous, regional metamorphic and contact rocks. Hornblende schists.

RIEBECKITE.

Composition: $nNaFeSi_2O_6$. $FeSiO_3$.
Criteria for determination in thin section:
Form: Similar to hornblende.

Optical Properties: Biaxial and negative.

Color: Dark blue to black.

Refringence same as hornblende.

Birefringence weak (0.005).   Interference colors masked by absorption.

Optic angle parallel to 010.   Angle large.

Pleochroism intense.   Z is yellowish green, Y is blue, X is indigo blue, nearly black.

Absorption marked.

Dispersion strong.

Differentiation: Characterized by intense color and pleochroism, strong dispersion and pronounced color.

Occurrence: In soda-rich igneous rocks and in some metamorphic rocks.

MICA GROUP.

Muscovite.

Sericite.

Biotite.

Lepidolite.

Phlogopite.

Criteria for determination in thin section:

Form: Monoclinic or pseudohexagonal.   As scales, which may be notched or jagged, with lateral sections lath-shaped.   As shreds, characterized by perfect basal cleavage giving thin laminæ.   Plates of hexagonal outline with the planes 001, 110, 010, with angles of 60 and 90 degrees.   Twinning common after the mica law in a plane perpendicular to 001 and practically parallel to 110.   Zonal structure common in the dark varieties. Elongation parallel to the cleavage and the *c* axis.

Optical Properties: Biaxial and negative.

Colors given under each variety.

Refringence medium.   Relief distinct.

Birefringence very strong, particularly in the colored micas, varying from 0.037 to 0.05. Interference colors of the third order, which may be very brilliant in thin sections of colorless mica, often appearing iridescent. Occasionally masked by strong absorption.

Extinction about parallel to the cleavage lines. Very small extinction angles may be noticed in biotite.

Absorption strong in colored micas.

Optic angle large in white micas and small in the ferro-magnesian varieties, appearing almost uniaxial.

Differentiation: Characterized by distinct relief, strong birefringence (chlorite has weak), one perfect cleavage marked by parallel, fine lines, practically parallel extinction, mottled appearance between crossed nicols, maximum extinction in colored varieties parallel to the vibration plane of the polarizer.

MUSCOVITE.

Composition: $H_2(K, Na) Al_3(SiO_4)_3$.

Criteria for determination in thin section:

See also under Mica group.

Colorless in thin section.

Inclusions: Not as common as in biotite. Zircon, apatite, spinel, garnet, quartz, and magnetite.

Alteration: By ordinary weathering to sericite, serpentine, talc.

Occurrence: Most common of the micas. Normal constituent of igneous rocks, especially granites. Abundant in gneisses and schists. Present in veins. Occurs as an alteration product of the feldspars, nephelite, etc.

Differentiation: From talc, by large optic angle.

From kaolinite, by strong birefringence.

From other micas, by being colorless in thin section.

From chlorite, by strong birefringence and lack of color.

SERICITE.

Sericite is a fine, scaly or fibrous variety of muscovite, with a greater degree of hydration. It is nearly uniaxial in character, with a small optic angle.

BIOTITE.

Pseudohexagonal.

Composition: $(K, H)_2 (Mg, Fe)_2 (Al, Fe)_2 (SiO_4)_3$.
Criteria for determination in thin section:
See also under Mica group.
Optical Properties:

Color: Black, green, brown, red, yellow.

Angle of optic axes almost 0 degrees in most biotite of igneous rocks.

Birefringence increases with increase in iron-content.

Absorption marked.

Pleochroism distinct. Z is dark to opaque brown, Y is the same, X is pure yellow.

Inclusions: Apatite and zircon common. Pleochroic halos abundant about inclusions.

Alteration: Reaily to chlorite, often accompanied by the formation of calcite, epidote and quartz.

Differentiation: From alkaline micas by small optic angle color and distinct pleochroism.

From chlorite, by strong birefringence and color.

From hornblende, by extinction parallel to cleavage and almost uniaxial interference figures in convergent light.

Occurrence: Important constituent of many igneous rocks, gneisses and schists. Developed by regional and contact metamorphism.

LEPIDOLITE.

Composition: $(Li, K) Al(F, OH)_2 Al(SiO_3)_3$.
Criteria for determination in thin section:

See also under Mica group.

Colorless in thin section.

Pleochroism distinct. Z and Y pink; X, colorless.

Differentiation: Microscopically   indistinguishable from muscovite. Pink color is believed to be due to traces of manganese.

Occurrence: In veins and dikes in granite, associated with cassiterite, tourmaline, etc.

Uses: As a source of lithium salts.

PHLOGOPITE.

Composition: $(K, H)_3 (Mg, F)_3 Mg_3 Al (SiO_4)_3$.

Criteria for determination in thin section:

See also under Mica group.

Color: Brown, brownish red, green, yellow.

Pleochroism, Z and Y are brownish yellow, X is colorless.

Inclusions: Hematite, rutile and tourmaline are common.

Differentiation: From minerals of other groups, same as biotite.

From muscovite, paragonite and lepidolite by color.

From biotite, by mode of occurrence.

Occurrence: Only in crystalline limestones, dolomites, and serpentines, associated with spinel, graphite, etc. Absent in igneous rocks.

Uses: As an insulator in electrical apparatus.

CHLORITE GROUP.
Penninite.
Clinochlore.

Character: Similar to the micas, with perfect cleavage parallel to 001, the basal pinacoid. This cleavage may not be noticed in fibrous or secondary chlorite. Aggregates of small, flat scales of irregular outline, usually with a laminated structure. Often in minute grains as a pigment in other minerals. Twinning common after the base and after the mica law.

Criteria for determination in thin section:

Color: Characteristically green, due to iron protoxide, varying from greenish white to dark green.

Refringence low. No relief. In penninite, n is 1.579 and 1.576. In clinochlore, n is 1.596 and 1.585.

Birefrigence usually weak, with interference colors of the low first order gray and bluish gray. Penninite (0.002). Clinochlore (0.011).

Extinction: Plates parallel to cleavage show at times isotropic characteristics. In other sections, extinction is apparently parallel to the cleavage. Clinochlore occasionally shows perceptible extinction angles.

Pleochroism present in all chlorites in green and yellow tints, the green being parallel to the cleavage.

Maximum absorption always in the direction of the cleavage.

Differentiation: From serpentine, by greater pleochroism. From mica, by weaker birefringence. Characteristics are pale green color, distinct pleochroism, low relief and weak birefringence.

Occurrence: Widely distributed, forming essential constituent of chlorite schist. Occurs secondary in igneous and metamorphic rocks, from the micas, amphiboles, pyroxenes, and garnets.

PENNINITE.

Pseudorhombohedral.

Composition: $H_8(Mg, Fe)_5 Al_2Si_3O_{18}$.

Differentiation from clinochlore: Nearly uniaxial character, negative sign, very weak birefringence, parallel extinction.

CLINOCHLORE.

Composition: Same as penninite.

Differentiation from penninite: Distinctly biaxial, positive sign, higher birefringence, occasionally oblique extinction, common polysynthetic twinning.

EPIDOTE GROUP.

Composition: $Ca_2(Al, Fe)_2(Al, OH)(SiO_4)_3$.

Criteria for determination in thin section:

Form: Columnar crystals, nearly always elongated, parallel to the b axis.   Fibrous, massive, or in irregular grains as aggregates.   Twinning common parallel to 100. Cleavage parallel to the basal pinacoid, imperfect parallel to the orthopinacoid.   Basal cleavage cracks not very numerous, and appear parallel to the general direction of elongation.

Optical Properties: Biaxial and negative.

Colorless to orange yellow in thin section.

Refringence high with rough surface. $n = 1.767$ and 1.730.

Birefringence variable, often strong, with high interference colors.   Variable in a single crystal (0.037).

Extinction parallel to cleavage in elongated sections.   In other sections, extinction angle varies from 0 to 28 degrees.

Interference figure of cleavage flakes show an axial bar with concentric rings.   Axial plane at right angles to the elongation of the crystal.   Axial angles are large.

Pleochroism: Z is colorless, yellowish green, pink; Y is pale blue to greenish yellow; X is colorless, lemon yellow, pale green.

Alteration: Epidote is very resistant to weathering.

Differentiation: From   light   colored   monoclinic

pyroxenes, by having optic plane at right angles to the principal cleavage cracks, which are parallel to the direction of elongation. Epidote is characterized by form and color, high refringence, parallel extinction in longitudinal sections, strong birefringence variable in a single crystal.

Occurrence: Very common, especially in schists and in zones produced by contact metamorphism between granites and limestones. Also as an alteration product of the ferro-magnesian minerals and feldspars in igneous rocks.

ZOISITE.

(Orthorhombic member of the Epidote group.)

Composition: Same as epidote without the iron.

Criteria for determination in thin section:

Form: Prismatic crystals or granular aggregates. Lamellar, fibrous or in compact masses. Perfect cleavage, parallel to 010; difficult, parallel to 100. Longer individuals show transverse parting. Microscopic twinning in polysynthetic bands occur.

Optical Properties: Biaxial and positive.

Colorless to yellow tints. Usually lacks color.

Refringence high, with rough surface. $n = 1.702$ and 1.697.

Birefringence weak, with grayish or whitish interference colors (0.005).

Extinction always parallel.

Differentiation: From epidote, by its lack of color and its weaker birefringence. It is characterized by parallel extinction, high relief, very weak birefringence, strong dispersion.

Occurrence: In crystalline schists, associated with amphibole, particularly hornblende. In igneous rocks, as an alteration of the feldspars. In veins in altered basic igneous rocks with quartz.

KAOLINITE.

Composition: $H_4Al_2Si_2O_9$.

Criteria for determination in thin section:

Form: Pseudohexagonal, in thin plates or scales. Usually in clay-like masses.

Optical Properties. Negative.

Colorless in thin section. Aggregates are cloudy. Refringence low with no relief. $n = 1.563$.

Birefringence weak (0.007).

Differentiation: From muscovite and talc by weak birefringence.

Occurrence: Kaolinite is the most common secondary mineral. It is derived from the feldspars by ordinary weathering. Occurs in large sedimentary clay masses as a result of the decomposition of aluminous silicates.

Uses: It is used in the manufacture of porcelain, pottery, and china.

TITANITE (SPHENE).

Composition: $CaTiSiO_5$.

Criteria for determination in thin section:

Form: In detached crystals and in disseminated grains. Often wedge-shaped when primary, and irregular grains when secondary. Flattened parallel to the base. Elongated parallel to a or c. Cleavage imperfect, parallel to the prism, appearing as a few rough cracks. Cleavage rarely observed in secondary forms. Twinning seen only between crossed nicols, the twinning boundaries bisecting the acute angles of the rhombs.

Optical Properties:

Colorless, brownish or yellowish.

Refringence high, with rough surface. $n = 2.009$ and 1.888.

Birefringence extremely strong, with interference colors of a high order, like those of calcite (0.1214).

Pleochroism strong in deeply colored varieties, appearing yellowish, parallel to *a* and reddish parallel to *c*.

Optic plane parallel to 010, with a small optic angle.

Inclusions: Often grouped about the center of the crystal.

Alteration: To a light yellow amorphous mass with calcite.

Differentiation: From staurolite, by the fact that the optic plane is in the shorter diagonal of the cross section instead of in the longer. It is characterized by high relief, extreme birefringence, biaxial character and positive sign.

Occurrence: Widely distributed as an accessory mineral in igneous rocks. Occurs in schists and gneisses. Is found as a secondary product called leucoxene, derived from ilmenite in basic igneous rocks.

FELDSPAR GROUP.

Monoclinic Feldspar.
Orthoclase.

Triclinic Feldspars.
Microcline.
Albite.
Oligoclase.
Labradorite.
Anorthite.

Composition: Silicates of aluminum with potassium, sodium, calcium, rarely barium.

Form: Crystal forms are similar, often short prismatic, somewhat flattened, parallel to 010. Narrow bands of albite, intergrown with orthoclase or microcline, forming "perthite" common. Plagioclase feldspars are triclinic, but angle alpha varies little from 90 degrees.

Cleavage: Perfect parallel to the basal pinacoid and almost as perfect parallel to the clinopinacoid. Cleavage cracks usually noticed only in very thin sections. The two cleavages intersect at 90 degrees in orthoclase and at 93 or 94 degrees in the plagioclase feldspars. The cleavage is not as distinct as the cleavage of mica or hornblende.

Twinning: Twinning common in the feldspars following the Carlsbad, Mannebach, Baveno, Albite and Pericline laws.

CARLSBAD TWINNING is the simplest type of feldspar twinning, and it occurs in both monoclinic and triclinic varieties, in the latter case causing confusion with other types of twinning. The twinning plane is the orthopinacoid and the composition face is the clinopinacoid. Carlsbad twins always consist of two individuals, a fact which may be used to differentiate between the plagioclase feldspars and orthoclase.

MANNEBACH TWINNING: The basal pinacoid is the composition face and twinning plane. This type of twinning is not common.

BAVENO TWINNING: The twinning plane is the clinodome to which the twinning axis is normal. Sections cutting such a twin show square or rhombohedral outlines, the cleavages being parallel to the sides.

ALBITE AND PERICLINE TWINNING are especially common on the plagioclase feldspars and are used as a means of identification. They are usually visible to the naked eye. In thin sections they appear as polysynthetic striations in narrow alternating light and dark bands, which extinguish alternately upon being rotated. In the albite type of twin, the twinning axis is normal to the clinopinacoid. Hence the lamellæ are parallel to the clinopinacoid and the striations are visible only on the basal pinacoid and the orthopinacoid.

In the pericline twinning, the twinning axis is parallel to *b*. Therefore, the pericline striations are visible on all faces of the crystal. It is obvious that if twinning occurs on the clinopinacoid, it must be of the pericline type. In thin section, the pericline twinning is visible in any section except a section cut parallel to the composition face.

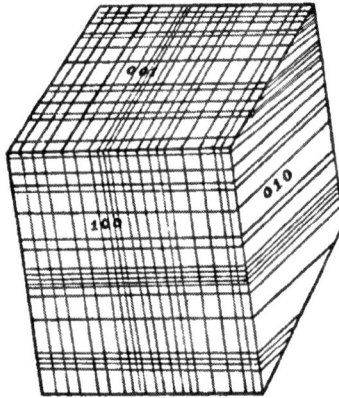

Fig. 19.   Triclinic feldspar form, showing the positions of the characteristic albite and pericline striations.

Differentiation of the Feldspars by Twinning.— Orthoclase occurs in simple twins, after the Carlsbad, Baveno and Mannebach laws, but never in polysynthetic twins.

Microcline is always polysynthetically twinned in two directions, a combination of albite and pericline twinning producing a rectangular crosshatching between crossed nicols.

Plagioclase feldspars practically always show a polysynthetic twinning, after the albite law.

Albite shows twinning lines that are fine and far apart, irregular and interrupted.

Oligoclase shows twinning lines that are clear and of regular widths.

Labradorite shows twinning lamellæ which are clear and definite, but the width often varies from one lamella to another.

Anorthite shows twinning lamellæ which are broad and regular, after the albite law, while those of the pericline law are distributed only in certain of the albite bands.

Optical Properties.—The optic plane containing the optic axes and the bisectrices is the chief optic element. Its position in each of the feldspars has definite relations to the cleavage, external faces, axes, and the positions of the albite twinning. In orthoclase and microcline, for example, the optic plane is almost parallel to the basal pinacoid, hence agrees with the direction of the most perfect cleavage. Z is perpendicular to the clinopinacoid. X lies in the plane of the clinopinacoid almost parallel to the base, varying about 5 degrees in microcline.

Refringence is low, similar to that of quartz. The Becke test is advised.

Birefringence is weak, similar to that of quartz.

TABLE OF REFRINGENCE OF THE FELDSPARS.

	n_g	n_m	n_p
Orthoclase . . .	1.526	1.5237	1.518
Microcline	1.5296	1.5264	1.5224
Albite	1.54	1.534	1.531
Oligoclase . . .	1.5469	1.5431	1.5389
Labradorite . . .	1.5625	1.5578	1.5548
Anorthite	1.5884	1.5837	1.5757

TABLE OF BIREFRINGENCE OF THE FELDSPARS.

	Section Normal to X $n_g - n_m$	Normal to Z $n_m - n_p$	Parallel to Optic Plane $n_g - n_p$
Orthoclase . . .	0.0023	0.0047	0.007
Microcline	0.0032	0.004	0.0072
Albite	0.006	0.003	0.009
Oligoclase	0.0038	0.0042	0.008
Labradorite . . .	0.0047	0.003	0.0077
Anorthite	0.0047	0.008	0.0127

Alteration: In zone of weathering to kaolinite, quartz, and calcite. The alteration of the feldspars to kaolinite or to other closely associated hydrous aluminum silicates is the ordinary method of origin of clay, and takes place more frequently in the acid than in the basic feldspars by a leaching out of the potash and hydration. The alteration begins along cleavage cracks, and finally spreads over the entire feldspar, causing it to appear opaque or cloudy. The kaolinite is usually in small flakes.

The alteration of the acid feldspars to sericite, a variety of muscovite, is common, the alteration taking place first along cleavage cracks. This is usually accomplished through the agency of hot solutions.

The basic feldspars frequently alter to chlorite, also to epidote associated with quartz and calcite.

Occurrence: Feldspars are the most abundant and the most widely distributed minerals of the earth's crust, occur abundantly in metamorphic rocks, and frequently in sedimentary rocks. Also in veins.

Differentiation: From quartz, by presence of cleavage and twinning, and biaxial character. Characterized by frequency of occurrence in practically all conditions, low refringence, and weak birefringence. Also by readiness with which they alter. To distinguish one feldspar from another the twinning, extinction angles, optic sign and refringence as determined by the Becke test are aids.

To distiguish one plagioclase feldspar from another, several practical methods have been devised.

1. EXTINCTION ANGLES ON BASE AND BRACHY-PINACOID.

Schuster established relations existing between the extinction angles on the base and the brachypinacoid. The prevalence of favorable cleavages aids in this determination. As these minerals are all triclinic, extinction takes place in all sections unsymmetrically with respect to crystallographic, twinning or cleavage lines. Consequently, extinction angles will always be observed. When the extinction angles on both the basal pinacoid and the brachypinacoid are large, anorthite is in all probability the mineral observed. When the angles are both small, the feldspar is oligoclase. Albite and labradorite show intermediate extinction angles. Orthoclase has extinction on the basal pinacoid of from 5 to 9 degrees.

The extinction angles given in the following table are marked plus or minus. The angles on the base and brachypinacoid are marked plus when the direction of extinction has apparently moved, as the hands of a watch, with reference to the upper right-hand edge of the crystal, between the base and pinacoid. The angles are marked minus when the reverse is true.

⸭EXTINCTION ANGLES.

Section parallel to base measured from trace of pinacoidal cleavage.		Section parallel to brachypinacoid measured from trace of basal cleavage.	
Albite	4	Albite	20
Oligoclase	2	Oligoclase	7
Labradorite	— 5½	Labradorite	— 20
Anorthite	— 37	Anorthite	— 42

2. STATISTICAL METHOD.

The method proposed by Michel Levy is practical in all sections, showing the albite twinning. This method consists in finding the maximum equal extinctions on opposite sides of an albite twinning line. The position of the plane which gives maximum extinctions in the zone normal to the brachypinacoid is different for different feldspars. This method, though tedious, is reliable, in that the various species have characteristic maxima.

Sections normal to the brachypinacoid may be recognized by the fact that the twinned parts show equal illumination eight times upon a complete rotation of the stage, once every 45 degrees, in which position, the two parts seem to belong to one individual. The faintly discernible twin line must be parallel to the plane of vibration of either of the nicols at equal illumination.

Maximum extinction angles in sections perpendicular to albite twinning:

Albite 16
Oligoclase 2
Labradorite 34
Anorthite ....Over 37

Monoclinic Feldspar.
ORTHOCLASE.

Composition: $KAlSi_3O_8$.
Criteria for determination in thin section:
See also under Feldspar group.

Twinning after Carlsbad law common, after Baveno and Mannebach less common.

Optical Properties: Biaxial and negative.
Colorless in thin section.
Refringence low. Relief absent and surface smooth.

Birefringence very weak, with interference colors of the lower first order, bluish gray, white, etc., not quite as bright as the colors of quartz and plagioclase.

Alteration to kaolinite so prevalent that surface usually appears cloudy.

Differentiation: From other feldspars, see under Feldspar group.

From quartz, by cloudy appearance, and negative character.

Occurrence: Abundant in acid plutonic rocks, present in intermediate and certain basic igneous rocks, in schists, gneisses, and in contact zones. As perthite, with bands of albite.

Uses: In the manufacture of porcelain and china. A variety of orthoclase called moonstone is used as a gem.

SANADINE.

Sanadine is a clear, glassy variety of orthoclase, occurring in rhyolite, trachyte, obsidian, etc. It decomposes less readily than orthoclase, has a smaller axial angle, and usually contains more inclusions.

Triclinic Feldspars.

MICROCLINE.

Composition: $KAlSi_3O_8$.

General characters same as orthoclase.

Differentiation: From orthoclase; simple crystals not showing the crossed twinning have extinction angles of about 15 degrees on the base with reference to the brachypinacoidal cleavage.

From other feldspars by the characteristic crossed, rectangular, grating structure.

Occurrence: Similar to orthoclase, but more abundant in pegmatites.

ALBITE.

Composition: $NaAlSi_3O_8$.

Form and cleavage characteristic of Feldspar group.

Optical Properties: Biaxial and positive.

See under Feldspar group.

Differentiation: From orthoclase, by the presence of polysynthetic twinning.

From microcline, by the absence of grating structure.

From other plagioclase feldspars, see under Feldspar group.

Occurrence: Seldom as a primary constituent of igneous rock except as an intergrowth with orthoclase or microcline in the form of perthite in soda-rich igneous rocks.

OLIGOCLASE.

Composition: Ab_6An_1 to Ab_1An_2.

Differentiation: See under Feldspar group.

Alteration: Kaolinization is less frequent than in albite.

Occurrence: In eruptive rocks and in crystalline schists.   More common in granites than albite is.

LABRADORITE.

Composition: Ab_1An_1 to Ab_1An_2.

Differentiation: See under Feldspar group.

Alteration: To a micaceous mass.   To an aggregate composed of zoisite, epidote, albite, quartz, etc.

Inclusions: Hematite and ilmenite abundant in colored varieties.

Occurrence: Common in basic igneous rocks as gabbro, basalt, etc., with olivine, augite and magnetite.   Is the principal constituent of anorthosite. Occurs sparingly in meteorites.

Uses: As an ornamental stone.

ANORTHITE.

Composition: $CaAl_2Si_2O_8$.

Differentiation: Anorthite has the strongest birefringence and the highest refringence of all of the feldspars. See under Feldspar group.

Occurrence: As an essential constituent of basic igneous rocks.

Developed by contact and regional metamorphism.

PART TWO.—PETROGRAPHY.

CHAPTER 8.

General Discussion of Igneous Rocks.

Petrography is that division of Petrology which is concerned with the systematic classification and description of rocks megascopically and microscopically.

The broad classification of rocks according to origin is: 1, Igneous; 2, Sedimentary, and 3, Metàmorphic.

Igneous rocks are those which have solidified by cooling from a molten condition.

Sedimentary rocks are those which have been deposited under water or on land by mechanical, chemical or organic processes.

Metamorphic rocks are those which are derived from previously existing igneous or sedimentary rocks by heat alone or by pressure and resultant heat.

Igneous Rocks.

Classification.—Several methods for classifying igneous rocks have been devised, two of which seem to be more or less satisfactory for practical purposes. These methods are the qualitative and the quantitative classifications. By an examination of the minerals comprising the rock many inferences may be derived as to the mode of origin, the conditions of crystallization, the general chemical composition, whether acid or basic, etc. A macroscopic observation alone gives the observer some basis for a rough classification. In the field, the mining engineer or geologist may find a rock which is essentially quartz and feldspar, with a small percentage of ferromagnesian minerals. He may call it a granite. His classification is correct if the rock contains 25 or 30 per cent of quartz. But if it contains only a few per cent, he will hesitate as to whether the rock is a granite or a syenite.

As long as such a doubt exists as to the proper classification of a rock, it is obvious that the system of classification is at fault. It is of course clear that all possible gradations in mineral percentages exist between the various igneous rock types. In so far as this is true, a qualitative classification based upon mineral percentages is defective. On the other hand, this method is exceedingly rapid in that one who is skilled in the manipulation of the microscope and in the interpretation of the phenomena observed in thin section can infer much from a glance about the nature of the rock.

For more complete descriptions of a rock, the quantitative classification is more satisfactory in that the chemical composition of the rock is used as a basis for classification. But such an analysis usually takes two or three days of careful work by a skilled chemist. Obviously, a qualitative classification with the aid of the microscope meets the requirements of the great majority of cases generally met with.

Essential and Accessory Minerals.—Of the thousand minerals which are known, only about ninety occur in igneous rocks. Twenty-five of these are of prime importance in determining the classification of a rock. These are the "essential minerals," for their presence is essential to the classification and definition of the rock type in which they appear. The remaining minerals, which comprise the majority, are the "accessory minerals," whose presence or absence does not influence the name under which the rock is classified. They are usually, though not always, present in small quantities. Typical accessory minerals are zircon, apatite, ilmenite, titanite, etc.

It is to be noted that minerals which are not essential to the definition of a large division, as the granite group, may become essential if the group is subdivided into a smaller division, as the amphibole-granite class.

Primary and Secondary Minerals.—Primary minerals are those which crystallized out from solution at the time of the solidification of the magma. Examples are the feldspars and quartz in granite. Secondary minerals are those which have formed after the solidification of the magma by the alteration of the previously existing minerals, the alteration usually taking place through the agency of weathering. Examples are the alteration of the feldspars to kaolinite and muscovite, and of the amphiboles and the pyroxenes to serpentine. Secondary minerals are derived from primary minerals by the application of heat and pressure. Sericites are thus derived from impure quartzites; chlorite from amphiboles and pyroxenes; talc from amphiboles, pyroxenes and impure dolomites.

Texture.—Although composition is the chief means of distinguishing rock types according to the qualitative system, texture likewise plays an important role in that it furnishes an important clue as to the circumstances under which the rock was formed. There are thus two considerations to be taken into account in the identification of a rock — the chemical composition, and the texture.

Texture is defined as the size, shape and mode of aggregation of the constituent particles of a rock. It is determined by certain conditions prevailing in and about the molten magma at the time of the solidification of the rock mass. Most important of these are:

1. The rate of cooling;
2. The chemical composition of the magma;
3. Pressure;
4. Temperature;
5. Action of mineralizers as steam, HCl, Fl, B.

The rock solidifying at great depths cools very slowly, allowing the minerals time to crystallize into well-formed individuals. Many minerals crystallize simultaneously, and these minerals interfere with each other as they grow. The interpenetration or irregular boundary line between any two crystals is a mutual adjustment of simultaneous formation. Molten magmas which are suddenly subjected to rapid cooling, such as would accompany an extrusion on or near the surface, crystallize relatively rapidly, with the result that a portion of the rock mass crystallizes as a glass. Microscopic crystals usually have time to make their appearance. It is also found that crystals of some minerals grow more rapidly than crystals of other minerals. When a rock shows a glassy appearance, with minute crystals embedded in the glass, the glass is regarded as a "groundmass."

The common textures may be reduced to four, as follows:

1. Glassy;
2. Felsitic, or stony;
3. Porphyritic;
4. Granitoid.

Glassy texture is characterized by absence of crystallization.

Felsitic or stony texture shows some crystallization of minute crystals enmeshed in a glassy or dense groundmass, giving the rock a stony or noncrystalline appearance.

Porphyritic texture results from conditions within the magma which allow the crystallization of certain minerals to take place before any other appears. These well-defined crystals, which are called "phenocrysts," are embedded in a finer ground mass, which may be wholly glassy, partly crystalline, or very finely crystalline throughout.

Granitoid texture is applied to those rocks which contain no groundmass and which are composed of crystals of the same general time of growth or which separated out in order of their basicity. In this case the earlier minerals show well-defined boundaries and crystal planes, whereas the later minerals fill the interstices and assume an irregular shape, determined by the position of the earlier minerals.

Extrusive flows and intrusive lavas and dikes are usually characterized by the presence of a groundmass. The deep-seated rocks are characterized by a granitoid texture. All gradations in texture between rocks consisting entirely of glass and of wholly crystalline material exist.

To illustrate the use of mineral composition and texture in classifying rocks, the following examples are given:

ROCK.	MINERALS.	TEXTURE.
Granite—	Alkali feldspar and quartz .	Granitoid.
Syenite—	same	Groundmass present.
Diorite—	Acid feldspars . . .	Granitoid.
Andesite—	same	Groundmass present.
Gabbro—	Lime feldspars . . .	Granitoid.
Basalt—	same	Groundmass present.
Diabase—	same but with intermediate texture.	

Textural Terms.—Convenient terms which are applied to igneous rocks to describe the amount of crystallized matter present, are:

1. Glassy, in which no crystals are present;

2. Cryptocrystalline, in which crystals are present but visible neither to the eye nor to the microscope;

3. Microcrystalline, in which crystals are present but visible only under the microscope;

4. Hypocrystalline, in which the rock consists partly of glass and partly of crystallized matter;

5. Holocrystalline, in which the rock is completely crystallized and no glass exists.

A classification of terms which describe the form and shape of the crystals:

1. Idiomorphic crystals are those which have their own peculiar geometric form. The first minerals to crystallize from any solution are idiomorphic, as they were allowed to grow without interference.

2. Hypidiomorphic crystals are those having part of their planes present and part absent. This may be brought about by an overlap in the time of crystallization of a series of minerals. One mineral is not given time to crystallize completely before an adjacent mineral interferes.

3. Allotriomorphic crystals are those in which no crystallographic planes are present. This is true of the last minerals to crystallize. Their shape is determined by previously existing minerals. Simultaneous crystallization sometimes results in the development of allotriomorphic crystals.

Rosenbusch's Law.—There is a normal order of crystallization in igneous rocks which in general is a law of

decreasing basicity, and is determined by the presence
or absence of silica, the chief acid radicle, in rock-form-
ing minerals. It was first worked out by Rosenbusch
(Heidelberg) and is briefly as follows:

1. Minor accessories: Apatite, magnetite, hematite,
ilmenite, pyrite, chalcopyrite, pyrrhotite, zircon, titanite,
garnet.

2. Ferro-magnesian minerals: Olivine, orthorhombic
pyroxenes, monoclinic pyroxenes, amphiboles, biotite,
muscovite.

3. Feldspathic minerals: Plagioclase feldspars in
order, from anorthite through bytownite, labradorite,
andesine, oligoclase, to albite; orthoclase, feldspathoids,
nephelite, leucite, sodalite. These latter minerals may
crystallize out before or after the feldspars.

4. Quartz, microcline. Quartz sometimes shows inter-
growths with orthoclase.

Volcanic and Plutonic Rocks.—Igneous rocks are
finally divided into two important types, which depend
directly upon mode of occurrence.

1. Volcanic or eruptive rocks are those which flow
out upon the surface or are ejected into the upper crust
of the earth near enough to the surface to assume a texture characteristic of rapid cooling. Volcanic rocks have
a glassy or porphyritic texture.

They may be either holocrystalline porphyritic or
hypocrystalline porphyritic.

2. Plutonic or intrusive rocks are those which do not
reach the surface except by subsequent erosion of the
overlying strata. They take on a texture characteristic
of slow cooling. Plutonic rocks are therefore holocrystalline and granitoid, occasionally porphyritic. They
may be distinguished from volcanic rocks by the absence
of groundmass.

Geological Occurrence.—The following table lists the commonly observed original structures of igneous rocks.

VOLCANIC ROCKS.
Extrusive.

1. Pyroclastic or fragmental deposits, as ash or tuff.
2. Volcanic necks.
3. Lava flows or sheets. Overflow from fissures.

VOLCANIC OR PLUTONIC ROCKS.
Intrusive.

4. Intrusive sheets or sills.
5. Bysmaliths.
6. Laccoliths.
7. Dikes.

PLUTONIC ROCKS.
Intrusive.

8. Bosses or stocks.
9. Batholiths.

Petrogeny.

Magma.—A magma is a fused rock mass in mutual solution. The essential feature of a solution is its tendency to become homogeneous. This tendency is produced by diffusion, convection currents, differences in temperature, sinking of fragments of superincumbent rocks, etc.

Magmas do not originate in the places where they are now observed. They move

1. In the zone of flow:

a. By rising gradually, like a bubble of air in water, with a flowage of the rocks above so as to allow passage;

b. By overhead stoping and absorption;

c. By assimilation.

2. In the zone of fracture:

a. By following the course of least resistance through whatever openings exist.

b. By overhead stoping.

Differentiation.—The possible causes of differentiation in a still fluid magma are gravity and differences in temperature, of which gravity is by far the more important. This accounts for the accumulation and concentration of magnetite along the lower border of a magma. It is the first mineral to crystallize. ·

Magmatic Stoping.—Marginal assimilation is one of the methods of magma advance through overlying rock formations. This method is effective chiefly in the early part of the magma's history and takes place at the main contacts and along a relatively limited surface.

According to the theory of magmatic stoping, each batholithic magma in its gradual advance upward through the overlying rocks engulfs large blocks of rock from the roof and walls. This process is facilitated by the shattering which it is believed accompanies an intrusion, due to unequal heating of the country rock along the contacts. These blocks are thus dissolved at depths forming a compound magma by assimilation. The average crust rock is more soluble in basic rocks than in acid.

Crystallization.—A eutectic is that proportion of two or more substances that has the lowest freezing point for those substances. Eutectic aggregates represent later products of crystallization because the first mineral to. crystallize is that which is in excess as compared with certain standard proportions. Thus, an intimate intergrowth of quartz and feldspar is a proof of simultaneous crystallization.

If a third substance were added to a eutectic proportion, it would lower the temperature so as to approach a ternary eutectic, unless the third substance were present to an amount less than one per cent, in which case it may be considered negligible. A mineral which crystallizes late has an appreciable effect on a eutectic proportion. One which crystallizes early, as apatite, has no effect. Thus the importance of an accessory mineral depends upon its solubility in a eutectic, although it is usually present in such a small amount that it may be disregarded.

On the other ·hand, the gases will be present throughout the crystallization of the magma. They tend to lower the temperature of crystallization to a greater extent than do the accessory minerals, and they aid the magma to solidify to a crystalline mass instead of to a glass.

In perfect isomorphism, A and B form mixed crystals in any proportion, so that there is a complete series of possible varieties between end members. Such a series is obtained between minerals which agree very nearly in molecular volume and crystalline elements. The albite-anorthite series is an example.

In imperfect isomorphism only certain mixtures are possible, as A with some B, or B with some A. The orthoclase-albite series is an example. Orthoclase may contain some albite, but no continuous series of mixtures connects pure orthoclase and pure albite.

Influence of Gases on a Magma.—Gases present in magmas are in a condition of unstable equilibrium, particularly at slight depths, liberating heat by reaction with each other. With decrease in pressure the reaction between gases increases. Thus it was observed in Hawaii that the temperature of a lava lake changed a few hundred degrees in temperature as the amount of gases passing through it increased or decreased. The temperature increased with increase in the amount of gases.

Relation Between Composition of Igneous Rocks and Magmas.—A magma has a composition differing from that of an igneous rock by the amount of material in the magma which escapes prior to crystallization. No analyses are available showing the relative amounts of water vapor with other gases. Iddings believes that 99.9% of all gases escaping from magmas consists of water vapor. Other gases are: CO_2, N_2, H_2S, O, HCl, H_2, SO_2, S, CH_4, Fl, B.

The more basic the rock is the greater quantity of gases it contains. The liberation of all of the gases in the outer seventy miles of the earth would double the amount of nitrogen and carbon dioxide in the atmosphere. Less than seventy miles of earth's crust during consolidation would yield all of the gases of the atmosphere. Chamberlain believes that the gases of the atmosphere have had that source. The same conclusion may be drawn with regard to the water of the hydrosphere, assuming that the average per cent of water in igneous rocks is 2.3.

Aids in the Determination of Igneous Rocks in Hand Specimens.—By a megascopic examination of an igneous rock it is sometimes possible to make a fairly good estimate as to what the rock is. Rocks with glassy or felsitic texture may easily be distinguished from granitoid or porphyritic rocks. By color it is possible to determine whether a rock is acid or basic, as the color is influenced by the amount of ferro-magnesian minerals. The rocks which are known to occur most commonly in nature should be given first consideration in examining the unknown rock.

The texture of the rock should be examined first. Having noted the presence or absence of a groundmass, the feldspars should be examined. Pink feldspar is usually orthoclase or microcline. If Carlsbad twinning can be observed, the mineral is probably orthoclase or microcline.

Plagioclase feldspars are usually white, gray or bluish gray, sometimes with a flashing blue surface. Albite twinning is often observed as a polysynthetic striation. Labradorite is of a darker blue or gray than the other feldspars. The feldspars are less transparent and glassy than quartz.

Quartz is recognized by its vitreous, fresh appearance and transparent quality. A rock containing quartz will contain neither leucite nor nephelite. If leucite or nephelite can be determined, quartz is absent. This fact is inherent in the chemical composition of the magma. A rock containing leucite or nephelite is too low in silica for any to be present as quartz in excess.

Dark minerals which appear in an orthoclase-microcline rock are biotite, hornblende, or augite. Biotite is determined by the flashing black surface due to the perfect cleavage plane. A knife blade may be used to test the softness and the ease of cleavage. It is more difficult to distinguish hornblende from augite in that they are both hard and not readily cleavable. In good crystals, augite shows an eight-sided cross section, whereas hornblende has a six-sided cross section.

The black minerals of a plagioclase rock are biotite and hornblende rather than augite. If the rock is mainly basic, has a dark color, and a dull, stony appearance, the black mineral is probably augite. Olivine occurs in basic lavas in clear, glassy greenish-yellow grains.

A dense, volcanic rock which shows a groundmass and visible quartz is either rhyolite or dacite. Without further examination the observer would be justified to call the rock rhyolite, as it is far more abundant than dacite.

If the volcanic rock is black and felsitic or stony in appearance, it is a basalt. If the rock answers neither of these descriptions but is evidently volcanic, it may be a trachyte, a phonolite, or an andesite. Of the three, andesite is the most probable, as it is the most common. It usually appears medium dark, midway between the acid and basic members of the series. If Carlsbad twinning is seen on the feldspar, the rock may be trachyte instead of rhyolite. If leucite is distinguished, it is a phonolite, otherwise there would be no justification for naming it thus.

A rock possessing a granitoid texture and quartz in some abundance may be called a granite rather than the rarer quartz diorite. If it contains orthoclase and no quartz, the observer would doubtless classify the rock either as a syenite or as a nephelite syenite, although the former would be the more probable. Diorites are darker than the syenites and may be inferred from this fact alone if the character of the feldspars cannot be determined. If plagioclase can be distinguished the classification is simplified. The latter may also show the dark or gray blue color of labradorite.

Diabase may be determined by a peculiar texture, commonly called diabasic texture, on account of its characteristic appearance. White plagioclase is intimately intergrown with augite crystals, the plagioclase having developed first, hence taking a lath-shaped texture. The interstices between the feldspars were later filled by the augite. It is dark and of medium grain.

The more basic rocks are determined by the total absence of quartz and feldspar, and the nature of the ferro-magnesian mineral comprising the greater part of the rock.

Table of Rock Classification — Igneous Rocks.

Essential Minerals	Alkali Feldspar and Quartz.	Alkali Feldspar With Little or No Quartz.	Alkali Feldspar, with Nephelite or Leucite.	Acid Plagioclase With or Without Quartz.	Basic Plagioclase With Pyroxene.	Basic Plagioclase With Alkali Feldspar.	Basic Plagioclase, Pyroxene, Nephelite or Leucite.	Non-Feldspathic.		
								Pyroxene, Nephelite or Leucite With or Without Olivine.	Pyroxene or Hornblende With Olivine.	Pyroxene or Hornblende With No Olivine.
Volcanic	Rhyolite Liparite Quartz porphyry	Trachyte	Phonolite Leucite phonolite	Dacite Andesite	Basalt Trap Diabase	Trachy-dolerite	Tephrite Basanite	Nephelinite Leucitite Nepheline basalt Leucite basalt	Limburgite	Augitite
Plutonic	Granite	Syenite	Nephelite syenite Leucite syenite	Diorite	Gabbro Norite	Essexite	Theralite Shonkinite	Ijolite Missourite	Peridotite	Pyroxenite Hornblendite

.

CHAPTER 9.

IGNEOUS ROCK TYPES.

Plutonic Rocks.

THE GRANITE FAMILY.

Mineralogical Composition.—Essential minerals: alkali feldspar and quartz. Common minerals: biotite, muscovite, amphiboles, pyroxenes. Accessory minerals: magnetite, apatite, zircon, titanite, garnet, tourmaline.

Texture.—Granitoid.

Character.—Granites are generally light in color, in shades of white, gray and pink, occasionally darker, due to an increasing amount of biotite, amphiboles or pyroxenes, in which case the rocks are liable to grade into the syenite or diorite families.

Microcline is more common in granites than in any other kind of a rock. The quartz may contain minute rutile needles or tiny cavities filled with gas bubbles. Biotite is the commonest dark silicate.

Varieties of Granites.—There are two varieties of granites. The most common type consists of the alkali-lime variety, and the rarer type is called the alkali-granite variety. The essential difference between the two types is that the alkali-lime variety grades toward and into the diorite family, and the alkali-granite variety grades toward and into the alkali syenites and finally into the nephelite syenites. The pyroxenes and amphiboles of the alkali-lime variety are more basic, and contain considerable amounts of magnesium and calcium. These minerals do not appear in the alkali granites, but are substituted by alkali pyroxenes and amphiboles, such as ægirite and riebeckite.

Classification:

GRANITE FAMILY.

Alkali-Lime Granites.—Containing alkali feldspar (orthoclase, microcline, albite, perthite) and quartz.

a. Granitite, with addition of biotite.

b. Amphibole granitite, with addition of biotite and amphibole.

c. Pyroxene granitite, with addition of biotite and pyroxene.

d. Granite, with addition of muscovite and biotite.

e. Amphibole granite, with addition of muscovite, biotite and pyroxene.

f. Pyroxene granite, with addition of muscovite, biotite and pyroxene.

g. Tourmaline granite, with addition of tourmaline.

Alkali Granites.—Containing alkali feldspar and quartz.

a. Alkali granitite, with addition of biotite.

b. Aegirite granite, with addition of ægirite.

c. Riebeckite granite, with addition of riebeckite.

d. Aplite, no subordinate mineral except possibly a little muscovite.

BORDER PHASES OF GRANITES.

Pegmatites.—A pegmatite is a border phase of a granite often observed on the edges of bosses and batholiths. They are usually very coarsely crystalline vein-granites. consisting of quartz, feldspar, muscovite, tourmaline, beryl, spodumene and others. Due to the immense size which the crystals attain, pegmatites are sometimes called "giant granites." The largest crystal of spodumene on record was found in the Etta tin mines of the Black Hills. This crystal measured thirty feet in height. Beryl crystals weighing over a ton have been recorded. Muscovite mica in sheets three feet in diameter are quite common. In pegmatites, the essential minerals of granites are not always present. Quartz and beryl, quartz and tourmaline, mica and quartz, feldspar and tourmaline, are all possible combinations.

Pegmatites are usually regarded as a late phase of the eruption which produced the granite. The common occurrence of such minerals as tourmaline, topaz and apatite in pegmatite leads to the suggestion that the influence of the rare elements fluorine and boron may have had some influence in effecting the coarse crystallization which so frequently exists.

Graphic Granite.—Graphic granite is a variety of a pegmatite which consists of a curious form of intergrowth of quartz and the feldspars in such a manner that the cross fracture of the vein rock exposes a cuneiform or wedge-shaped texture resembling the writing character of the ancient Chaldeans and Assyrians. The most common intergrowth is quartz inclosed in orthoclase, microcline, or perthite. Since neither of the minerals comprising graphic granite possesses any definite crystal shape, it is evident that they crystallized from solution at the same time.

Greisen.—Greisen is another border phase of a granite mass which, although not occurring abundantly, is of economic value the world over as the mother rock for cassiterite, the tin ore. It is a granitoid rock composed of quartz and muscovite, or some related white mica, as lepidolite or zinnwaldite.

Greisens are the result of contact action on granites under the influence of mineralizers. It has been suggested that the water and fluorine vapors came into contact with the feldspars, converting them to micas.

Orbicular Granite.—Orbicular granite consists of spherical or ellipsoidal masses of basic minerals in granite. The dark minerals are commonly biotite, pyroxenes, or amphiboles. They exist in the nature of basic segregations.

Contact Metamorphism.—Owing to the amount of heat and the presence of mineralizers which are given off from the granitic magma, considerable metamorphic effect is exerted upon adjacent rocks with which the intruding magma comes in contact. This effect may not be conspicuous upon igneous and previously metamorphosed rocks, as they are already in a highly crystalline condition. The effect upon sedimentary rocks is usually considerable. It takes the form of a more or less complete recrystallization of the secondary and undecomposed primary minerals composing the rocks. In addition to complete recrystallization there is quite frequently an addition or a subtraction of constituents or an exchange of constituents with the magma.

The chief effect upon sandstone is a recrystallization of the rounded grains of sand, resulting in a filling of the interstices between the grains. Quartzite is the rock developed. If the sandstone contained considerable clay and other impurities, a sericite schist would develop.

A common effect of an intrusive granite in contact with a limestone formation is the recrystallization of the limestone to a marble. Most limestones contain considerable percentages of argillaceous and siliceous material, which will be converted to lime magnesium silicates. If there is no opportunity for the carbon dioxide of the limestone to escape, it will be retained in the marble as calcite or dolomite. If there is an opportunity for its release through fissures or joint cracks, the resulting mass may consist essentially of secondary silicates, some of which develop by a recrystallization of the original constituents of the limestones and others by the addition of material from the magma.

The contact metamorphic effect of an intrusive magma on shale is pronounced and characteristic. Immediately at the contact, the shale is converted into a "hornfels" rock, which is dense, very finely crystalline, extremely hard, has a conchoidal fracture and consists chiefly of quartz, feldspar and biotite. From hornfels to the unaltered shale the following stages are often observed: highly metamorphosed mica schist, spotted mica slate, spotted slate lacking the conspicuous development of the micas, unaltered shale. This change may be almost imperceptible, and may extend for miles from the actual contact. The chemical composition of hornfels and shale are often very similar, although at times it shows an addition of material in the hornfels.

Economic Uses of Granites.—Granite is more extensively used for structural purposes than any other igneous rock, although any crystalline rock is often loosely called granite in the quarry if it consists of silicates. Granite is the strongest of the common building stones, the crushing resistance ranging from 15,000 to 30,000 pounds per square inch tested on two-inch cubes.

The following resistance tests show the average in range:

Granite from St. Cloud, Minn., 26,250 to 28.000.
Granite from Mystic River, Conn., 18,125 to 22,250.
Granite from Cape Ann, Mass., 19,500.
Granite from Vinal Haven, Maine, 25,700.

Upon the following points are based the desirability of granites for structural purposes:

1. Homogeneity of texture.
2. Adaptability to tool treatment.
3. Good rectangular jointing in the quarry.
4. Pleasing color.
5. Transportation facilities.
6. Durability as affected by grain and mineral-content.

A light color is generally more desirable than a dark one, and a medium grain is more favorable for durability than a coarse grain. The Rapakiwi granite of southern Finland is used freely in Petrograd for columns. It contains large red orthoclase crystals, which give the rock a prevailing red color, greenish plagioclase, smoky quartz and biotite. The disintegration is found to be rapid, as the jointing or fracture occasioned by the cleavage planes of one mineral tends to continue into the others.

Relationship.—Granite approaches syenite by insensible gradations with decrease in quartz. It approaches diorite with increase in hornblende or biotite and plagioclase. Intermediate varieties are called granodiorites. With increase in augite and plagioclase, granite approaches gabbro.

Geographical Distribution.—Granite occurs abundantly along the Atlantic Coast from Virginia into Canada. It is extensively quarried. The Quincy granite from Quincy, Mass., is a well-known building granite. In Minnesota, Wisconsin, and northern Michigan, and northward in Ontario, much of the Pre-Cambrian crystalline area known geologically as the Laurentian Highland is granite. It is found widespread in the West, existing in the Black Hills, in the Wasatch, the Rocky and the Sierra Mountains.

Analyses of Granites:

	1	2	3	4	5	6	7
SiO_2 .	73.23	77.50	69.00	67.70	61.90	69.46	66.84
Al_2O_3 . .	15.47	10.10	14.80	14.80	13.20	17.50	18.32
Fe_2O_3 .	....	....	2.30	2.10	3.60	2.30	2.27
FeO . .	3.34	2 70	.90	3.40	2.30	....	.20
MgO . .	.24	.60	1.10	1.60	4.60	.30	.81
CaO . .	.80	2.30	3.80	3.90	3.50	2.70	3.31
Na_2O . .	1.70	3.20	2.50	4 10	2.70	2.93	5.14
K_2O . .	4.38	4.00	4.50	4.30	6.10	4.07	2.80
H_2O . .	.65	.3	.7	1.00	1.10	.82	.46
Total . .	99.81	100.70	99.60	102.00	99.00	99.95	100.49
Sp. Gr. .	....	....	2.68	2 62	2.72	2.68	....

1. Granite from Carlsbad, Bohemia, Austria.
2. Granite from Baveno, Lake Maggiore, Italy.
3. Granite from Barr, Lower Alsace, Germany.
4 Amphibole granitite from Barthoga, Sweden.
5. Pyroxene granite from Lavelline, Vosges Mountains, France.
6. Alkali granite from Chester, Massachusetts.
7. Augite soda granite from Kekequabic Lake, Minnesota.

Discussion of Analyses:

1. Granite contains more silica than any other plutonic rock.

2. Alumina content is not as high as in the syenites. It is present chiefly in feldspars and biotite.

3. Iron content is generally low. It is present chiefly in biotite, amphiboles, pyroxenes, and possibly magnetite.

4. Magnesia content is low, indicating an absence of many ferro-magnesian minerals.

5. Lime content is low. It is present chiefly in a few acid plagioclases, in amphiboles and pyroxenes.

6. Potash predominates over soda, occurring in alkali feldspar and biotite.

7. Soda rarely predominates over potash. When it does (see Analysis 7), albite is the chief feldspar. It marks a gradation toward the diorites.

8. A high water content is probably due to the formation of secondary minerals, as serpentine, kaolinite, etc. Water in small quantities exists in many primary minerals: micas, amphiboles.

9. The presence of apatite as an accessory mineral is indicated by the presence of P_2O_5.

10. The darker granites have the higher specific gravities.

11. Granite is similar to rhyolite in composition.

12. The alkali-lime granites contain more iron, magnesia, and lime than do the alkali granites. The alkali granites are richer in soda, potash, and possibly silica.

THE SYENITE FAMILY.

Mineralogical Composition.—Alkali feldspar with little or no quartz.

Texture.—Granitoid with groundmass absent. Sometimes porphyritic.

Character and Distribution.—Syenites are allied with nephelite syenites, into which they grade with increase of soda. They merge into the diorites with increase in plagioclase. Their geological occurrence is practically the same as that of granites. They are often found at the rims of granite bosses or batholiths where there has been a decrease in silica in the form of quartz.

Geographically the syenites form the basement of the White Mountains, and occur in dikes near Little Rock, Arkansas. Many minor occurrences have been recorded.

VARIETIES.

Alkali-Lime Syenite.—Essential minerals are alkali feldspar, with a little basic plagioclase.

a. Amphibole syenite, with addition of amphibole.

b. Mica syenite, with addition of mica.

c. Pyroxene syenite, with addition of pyroxene.

Alkali Syenite.—The alkali syenites are rare.   The same three types occur in this group as occur in the alkali-lime group, except that the dark minerals are alkali pyroxenes and amphiboles.

Belonging to the alkali-lime group is a rock called a "monzonite," which grades over into the diorites, as it contains both alkali feldspar and plagioclase.   A rock associated with the copper ore deposits of Butte, Montana, is a more acid rock of this type, called a quartz monzonite.   It covers an area seventy miles by forty, and is known as the Bowlder Batholith.

The corundum syenites north of Kingston, Ontario, are alkali syenites composed of pink orthoclase, and a greenish corundum which is used as an abrasive.

The colors of the syenites are light, although usually darker than the granites.   The crushing strength is greater.

Analyses of Syenites:

	1	2	3	4	5	6
SiO_2 . . .	64.00	51.00	59.40	52.88	59.78	59.83
TiO_2	....	1.80	.30	....	....	....
Al_2O_3 . .	17.40	14.50	17.90	20.30	16.86	16.85
Fe_2O_3 . . .	1.00	4.20	2.00	3.63	3.08	....
FeO . .	2.30	4.40	6.80	2.58	3.72	7.01
MgO . . .	.60	8.20	1.80	.79	.69	2.61
CaO . . .	1.00	5.10	4.20	3.03	2.96	4.43
Na_2O . . .	6.70	1.80	1.20	5.73	5.39	2.44
K_2O . . .	6.10	7.20	6.70	4.50	5.01	6.57
H_2O . . .	1.20	1.00	.40	1.01	1.58	1.29
P_2O_5 . . .	....	.70	.60	.54	....	....
Total . .	101.40	99.90	101.30	100.99	99.07	101.03
Sp. Gr. . .	....	....	2.77	2.73	2.67	2.73

1. Mica syenite (alkali type) from Tonsenoos, near Christiania, Norway.

2. Mica syenite (alkali lime type) from Gangenbach, Black Forest, Germany.

3. Hornblende syenite from Biella, Piedmont, Italy.
4. Augite syenite from Byskoven, near Laurvik, Norway.
5. Syenite from Custer County, Colorado.
6. Hornblende syenite from Plauenschen Grund, near Dresden.

Discussion of Analyses:

1. Silica content is lower than in granites, due to decrease of quartz.

2. Alumina content is higher than in granites, due to relative increase in feldspars.

3. Iron, magnesia and lime contents are higher, due to increase in ferro-magnesian minerals, chiefly hornblende.

4. Alkali content is higher, due to increase of feldspars.

5. The high water content is due to hydration of the secondary minerals.

6. The specific gravity is higher in that of granites, due to the increase in ferro-magnesian minerals.

7. The alkali-lime syenites contain more lime and magnesia than the alkali syenites.

NEPHELITE AND LEUCITE SYENITES.

Mineralogical Composition.—Alkali feldspar and nephelite or leucite.

Texture.—Granitoid, sometimes porphyritic.

Character and Distribution.—Nephelite and leucite syenites are white to smoky gray in color, and contain very few accessory minerals. When present, they usually are biotite, ægirite, and an alkali amphibole called barkevikite.

These types are comparatively rare, occurring especially as dikes. They are known in North America at

Montreal and Dungammon (Ontario), Litchfield (Maine), Red Hill (New Hampshire), Salem (Massachusetts), Beemersville (New Jersey), and near Little Rock (Arkansas), in well-known exposures, though they have a widespread occurrence.

Of economic importance is the occurrence of rare mineral containing zirconium, tantalum, titanium, yttrium, cerium, lanthanum, terbium, and other rare elements. In the nephelite-syenite pegmatites of southern Norway about 800 of these rare minerals have been recorded. A corundiferous nephelite syenite is found in commercial quantities in Canada.

Analyses of Nephelite and Leucite Syenites:

	1	2	3	4
SiO_2	56.30	50.36	60.39	50.90
Al_2O_3	24.14	19.34	22.51	19.67
Fe_2O_3	1.99	6.94	.42	7.76
FeO	....	....	2.26	....
MgO	.13	....	.13	.36
CaO	.69	3.43	.32	4.38
Na_2O	9.28	7.64	8.44	4.38
K_2O	6.79	7.17	4.77	6.77
H_2O	1.58	3.51	.57	1.38
Total	100.90	98.39	99.81	100.01

1. Nephelite syenite from Ditro, Transylvania, Hungary.
2. Nephelite syenite from Beemersville, Sussex County, New Jersey.
3. Nephelite syenite from Litchfield, Maine.
4. Leucite syenite from Magnet Cove, Arkansas.

Discussion of Analyses:

1. Silica content is lower than in the syenites, as nephelite has 44% silica, and the minerals which it replaces have several per cent more.

2. Alumina content is higher than in any other plutonic rock, due to the presence of nephelite or leucite.

3. Iron content is variable, but magnesia and lime contents are lower than in the syenites, due to the absence of many ferro-magnesian minerals.

4. Alkali content is higher than in any other plutonic rock. Soda predominates over potash in the nephelite syenites. In the leucite syenites potash increases, but may not predominate, as leucite readily decomposes, allowing the potash to be removed.

5. The specific gravity is less than that of the syenites.

DIORITE FAMILY.

Mineralogical Composition.—Acid plagioclase with or without quartz, and some dark mineral, most commonly an amphibole near green hornblende. Biotite is common.

Texture.—Granitoid, at times porphyritic.

Character.—The color of diorite is dark, due to the ferro-magnesian minerals. It grades from the alkali-lime granite type by decrease in alkali feldspar and increase in acid plagioclase. Certain intermediate phases are called granodiorites, containing both alkali feldspar and acid plagioclase.

Diorites are not very common in North America. They occasionally form on the edge of granite bosses or batholiths.

CLASSIFICATION OF THE DIORITES.

With Quartz	Without Quartz
Quartz mica diorite.	Mica diorite.
Quartz hornblende diorite.	Hornblende diorite.
Quartz augite diorite.	Augite diorite.
Quartz hypersthene diorite.	Hypersthene diorite.

Analyses of Diorites:

	1	2	3	4	5
SiO$_2$ . . .	61.22	64.12	56.09	52.45	52.00
Al$_2$O$_3$ . . .	16.14	16.50	16.03	18.63	15.75
Fe$_2$O$_3$ . . .	3.01	2.71	3.12	11.40	3.55
FeO	2.58	4.26	4.77	1.19	12.84
MgO . . .	4.21	2.34	8.03	5.16	3.42
CaO . . .	5.46	4.76	6.73	6.84	7.39
Na$_2$O . . .	4.48	3.92	3.49	2.64	2.37
K$_2$O	1.87	1.92	1.87	.37	1.24
H$_2$O . .	.44	.73	.16	2.40	.35
Total . . .	99.41	101.26	100.13	100.82	99.91

1. Pyroxene amphibole biotite diorite. Electric Peak, Yellowstone Park.
2. Quartz mica hypersthene diorite from Pfundersberg, Tyrol.
3. Mica hypersthene diorite from Campomaior, Portugal.
4. Amphibole diorite from Neunseestein Barr, Alsace.
5. Augite diorite from Richmond, Minnesota.

Discussion of Analyses:

These analyses when compared with the analyses of granite show that—

1. Iron, lime, magnesia and alumina contents are higher.

2. Alkali content is lower.

3. Soda predominates over potash.

4. Silica is lower, due to change of feldspar.

5. The specific gravity is higher.

GABBRO AND NORITE FAMILY.

Mineralogical Composition.—Basic plagioclase and usually a pyroxene.

Texture.—Granitoid. Never porphyritic.

Relationship.—Gabbros grade by decrease in plagioclase to the more basic pyroxenites and peridotites.

Varieties.—Essential to all, basic plagioclase.

1. Gabbro, with addition of diallage.

2. Hornblende gabbro, with addition of hornblende.

3. Olivine gabbro, with addition of olivine and diallage.

4. Norite, with addition of hypersthene, bronzite or enstatite.

5. Olivine norite, with addition of hypersthene, bronzite or enstatite and olivine.

6. Anorthosite, composed chiefly of labradorite. It may contain a few dark minerals which, when metamorphosed, cause the development of almandite garnets in considerable quantities.

Concentration of Magnetite.—The concentration of magnetite in many gabbroid magmas took place during the process of solidification along the lower border of the magma. This concentration was effected by the early crystallization of the magnetite from solution, its high specific gravity, convection currents, etc. Magnetite of this occurrence has been found in commercial quantities in the Adirondacks and in Lake and Cook counties of northern Minnesota. It is usually titaniferous, due to an intimate association with the mineral ilmenite.

Nickeliferous Pyrrhotite.—In the Sudbury district of Ontario, great quantities of nickeliferous pyrrhotite and workable amounts of chalcopyrite are found in the norite. They occur as magmatic segregations. The pyrrhotite is an important source of nickel.

In Lancaster County, Pennsylvania, nickeliferous pyrrhotite is observed along the contact of a metamorphosed basic igneous rock called amphibolite.

Platinum is believed to occur minutely disseminated in rocks of this type, the weathering of which has supplied the placer deposits.

Analyses of Gabbros and Norites:

	1	2	3	4	5
SiO_2 . . .	54.47	44.10	46.70	49.10	49.95
Al_2O_3 . . .	26.45	24.50	22.20	21 90	19.17
Fe_2O_3 . . .	1.30	7.90	.80	6.60	4.72
FeO . . .	.67	6.50	5.50	4.50	6.71
MgO . . .	.69	3.80	10.30	3.00	5.03
CaO . . .	10.80	12.00	11.70	8.20	9.61
Na_2O . . .	4.37	1.70	1.70	3.80	3.13
K_2O . . .	.92	.20	.10	1.60	.74
H_2O . .	.53	.60	1.10	1.90	.09
Total .	100.20	101.30	100.10	100.89	99.84
Sp. Gr. . . .	2.72	3.04	3.02	2.94	2.94

1. Anorthosite from Adirondacks, New York.
2. Gabbro from Mount Hope, Baltimore, Maryland.
3. Olivine gabbro from Langenlois, Lower Austria.
4. Hornblende gabbro from Duluth, Minnesota.
5. Norite from Monsino, near Iorea, Piedmont, Italy.

Discussion of Analyses.—The analyses compared with analyses of diorites show:

1. A lower silica content due to the absence of quartz and the decreasing basicity of the feldspars.

2. Higher alumina, lime, iron and magnesia content. High magnesia, as in Analysis 4, suggests olivine.

3. Lower alkali content.

4. Higher specific gravity.

ESSEXITE FAMILY.

Mineralogical Composition.—Basic plagioclase, with varying amounts of subordinate orthoclase. Nephelite or sodalite may be present. The dark minerals are augite, biotite, and a brown amphibole called barkevikite. Olivine and apatite sometimes occur. The plagioclase is usually labradorite, rarely andesine.

Relationship.—Essexite is related to the gabbros much as monzonite is related to the syenites. It was originally classed with the gabbros, but is more generally associated with the alkali and nephelite syenites. It may be considered intermediate between this type and the gabbroid type. It was first recognized in association with nephelite syenites near Boston.

Discussion of Analyses (See next table) :

1. Low silica content.
2. High alumina and iron content.
3. Equal amounts of lime and the alkalies.
4. Soda predominates over potash.
5. Magnesia content low as compared with gabbro.
6. P_2O_5 high due to apatite.

THERALITE-SHONKINITE-MALIGNITE FAMILY.

Mineralogical Composition.—Basic plagioclase, nephelite or leucite, pyroxene with some biotite, and rarely amphibole. Members of the sodalite group may accompany the nephelite.

Relationship.—These rocks are related to essexite, and grade into them. The presence of nephelite or leucite is the essential difference. Chemically they are quite similar.

Distribution.—Theralite was first found in the Crazy Mountains, Montana, near Livingston, and described by J. E. Wood, of Harvard.

Shonkinite was described by Weed and Pirrson, from Square Butte in the Highwood Mountains of Montana, as a border phase of a sodalite syenite laccolith. It contains little nephelite, but has instead sanadine. Consequently, potash prediminates over soda. In theralite, soda predominates.

Malignite was named by Lawson, from Puba Lake, Ontario. It contains chiefly ægirite, augite, biotite, orthoclase, nephelite and titanite.

Discussion of Analyses:

1. Chemical resemblance to essexite.
2. Low silica.
3. Equal lime and magnesia content.
4. Magnesia higher than in essexite.

Analyses:

	1	2	3	4	5	6
SiO₂ . . .	47.94	43.17	46.73	47.85	42.79	46.06
Al₂O₃ . . .	17.44	15.24	10.05	13.24	21.59	10.74
Fe₂O₃ . .	6.84	7.61	3.53	2.74	4.39	3.17
FeO . . .	6.51	2.67	8.20	2.65	2.33	5.61
MgO . .	2.02	5.81	4.68	5.68	1.87	14.74
CaO . . .	7.47	10.63	13.22	14.36	11.76	10.55
Na₂O . .	5.63	5.68	1.81	3.72	9.31	1.31
K₂O . . .	2.79	4.07	8.76	5.25	1.67	5.14
H₂O . . .	2.04	3.57	1.24	2.74	.99	1.44
P₂O₅ . . .	1.04	....	1.51	2.42	1.70	.21
Total . .	99.02	98.45	99.73	100.65	98.40	98.97

1. Essexite from Salem Rock, near Boston.
2. Theralite from Martinsdale, Crazy Mountains, Montana.
3. Shonkinite from Square Butte, Highwood Mountains, Montana.
4. Malignite from Puba Lake, Rainy Lake District, Ontario.
5. Ijolite from Iwaara, Finland.
6. Missourite from Shonkin Creek, Highwood Mountains, Montana.

IJOLITE AND MISSOURITE.

Mineralogical Composition.—Ijolite contains ægirite-augite and nephelite, often with apatite, titanite, and andradite as accessories. It is nonfeldspathic.

Missourite contains augite, leucite, olivine, and biotite with accessories. It is nonfeldspathic.

Relationship.—These rocks are end products of the series beginning with essexite. They are closely related to the theralite-shonkinite rocks and are distinguished from them by the fact that they contain no feldspars. Ijolite was first found on Mount Iwaara in northern Finland.

'   PERIDOTITE FAMILY.

Mineralogical Composition.—Olivine with pyroxenes, amphiboles, or biotite. No feldspars are present.

Relationship.—The peridotites grade from olivine gabbros by the elimination of the feldspars. They are found on the edges of gabbro and norite bosses. They are regarded as ultra basic.

Classification.—Olivine is essential in all varieties.

1. Sherzolite, by addition of diopside and enstatite.
2. Harburgite, by addition of enstatite.
3. Wehrlite, by addition of diallage and hornblende.
4. Cortlandite by addition of hornblende.
5. Dunite chiefly olivine. It may contain chromite or chrome spinel.
6. Kimberlite, which was named from its occurrence in Kimberley, South Africa, is a peridotite found in the truncated cones of extinct volcanoes. In its type locality it weathers to a soft serpentine rock called "blue ground." It is the mother rock of the diamond. A similar rock has been found in southern Arkansas, where diamonds are likewise found in commercial quantities. Small diamonds have been found in a peridotite rock in Elliot County, Kentucky.

Garnierite, the chief ore of nickel, is a secondary mineral associated with serpentinized peridotite, probably as an alteration of a nickel-bearing olivine. The French locality of New Caledonia is the only important locality.

Analyses:

	1	2	3	4
SiO_2	41.44	34.98	53.98	44.01
Al_2O_3	6.63	10.80	1.32	11.76
Fe_2O_3	13.87	1.42	1.41	15.01
FeO	6.30	21.33	3.90	....
MgO	18.42	19.30	22.59	25.25
CaO	7.20	.43	15.49	4.06
Na_2O	.24	.17	....	....
K_2O	.93	5.42	....	....
H_2O	5.60	1.28	.83	....
Total	100.63	95.13	99.59	100.09
Sp. Gr.	....	3.276	3.301	....

1. Amphibole peridotite from Sebreizheim, Baden, Germany.
2. Mica peridotite from Kaltes Thal, Harzburg, Germany.
3. Pyroxenite from Baltimore, Maryland.
4. Pyroxenite from Meadowbrook, Montana.

Discussion of Analyses:

1. Lower silica and alumina content than in gabbro, due to the absence of feldspars.

2. Iron content varies, depending upon the dark mineral present.

3. Magnesia content higher than in any other normal plutonic rock.

4. Lime content varies.

5. Alkali content less than that of any other igneous rock. In a mica peridotite, potash predominates over soda, an unusual case among alkali-lime rocks.

6. Specific gravity highest of the normal plutonic rocks.

PYROXENITE AND HORNBLENDITE FAMILY.

Mineralogical Composition.—These rocks consist of a single pyroxene, or a single amphibole, or two or more minerals of the same group.

Relationship.—They are the end products of the ultra basic rocks grading from the peridotites by the elimination of the olivine. The varieties depend upon the mineral which composes the rock, the rock name usually being the mineral name with the suffix "ite" added to it.

Varieties of the family are diallagite, enstatite, bronzitite, hypersthenite, hornblendite.

Occurrence.—The pyroxenites are usually found in association with gabbro and norite masses. Peridotites may have a similar occurrence. The serpentine deposits of Quebec and New England occur in this association. Serpentine asbestos is extracted in commercial quantities.

CHAPTER 10.

IGNEOUS ROCK TYPES.

Volcanic Rocks.

THE RHYOLITE FAMILY.

Mineralogical Composition.—Orthoclase, oligoclase, quartz. Biotite, hornblende. The rhyolites are chemically the equivalents of the granites, particularly the alkali-lime type.

Texture.—Well developed groundmass, often largely glassy. Frequently porphyritic.

Relationship.—Rhyolite grades imperceptibly into trachyte, granite, and dacite. Unless quartz is recognized, a microscopic examination is necessary to distinguish rhyolite. It may easily be confused with dacite unless the polysynthetic twinning of the feldspar characteristic of dacite can be seen.

Character.—The term "rhyolite" comes from the Greek verb *rhein*, "to flow," because of the flow structure frequently observed. Liparite is a synonymous term used largely in Europe. It was named from the Lipari Islands, in Sicily. Quartz porphyry is a term often applied to the rhyolites which have crystallized as intruded sheets, laccoliths, dikes, and sills. The glassy portion is characterized by its behavior between crossed nicols, remaining dark during a complete revolution of the stage.

The processes of weathering of the rhyolites are the same as take place in granites, ordinary decomposition by atmospheric agencies giving rise to the formation of the hydrous aluminum silicates. Metamorphic processes develop schistose textures leading in extreme cases to the development of sericite schists.

Early in the study of volcanic rocks it was customary to distinguish two types — those which had erupted previous to Tertiary times, and those which had erupted after Tertiary times. The former were called Paleovolcanic, and the latter were called Neovolcanic. Fortunately, this classification did not survive.

Classification.—Rhyolites are regarded by some writers as porphyritic rocks with phenocrysts of quartz and alkali feldspars in a groundmass which is wholly glassy or a very finely crystalline aggregate of quartz and feldspar. They classify in the "glasses" all varieties of volcanic rocks in which chilling has prevented crystallization.

The classification here adopted combines the glasses with the rhyolites.

Volcanic glasses are obsidian, pumice, pitchstone, and perlite.

OBSIDIAN is a dense, homogeneous glass with a low percentage of water.

PUMICE is a cellular glass formed by the expansion of the cooling magma by the escaping steam bubbles. It is light, very porous, and may resemble blast-furnace slag.

PITCHSTONE is essentially the same as obsidian, with a higher percentage of water. It is more resinous in appearance, giving it a greasy or pitchy luster.

PERLITE is a pitchstone which has a spheroidal arrangement of the particles, giving rise to a rounded fracture.

Pantellerite.—Pantellerite is a volcanic rock corresponding to the alkali granites. It is rare, and occurs so far as known only on the island of Pantellerea, in the Mediterranean Sea. It contains a rare feldspar called anorthoclase, which is an isomorphous mixture of albite and orthoclase.

Distribution.—Rhyolites are widespread throughout the Western States. Obsidian Cliff in Yellowstone Park, Silver Cliff in Utah, extinct volcanoes in New Mexico, Utah, Montana and California (Mono Lake), are well-known examples. In Leadville, Colorado, they are associated with the ore deposits.

Along the Eastern Coast, remnants of rhyolite lavas from ancient Pre-Cambrian volcanoes have been found in New Brunswick, Maine, Massachusetts, and Pennsylvania.

Analyses:

	1	2	3	4
SiO_2	83.59	77.00	75.60	68.30
Al_2O	5.42	12.80	11.50	10.90
Fe_2O	....	1.90	2.40	3.70
FeO	....	....	....	.40
MgO	....	.30	....	.20
CaO	3.44	1.40	.80	1.40
Na_2O	5.33	3.00	2 90	7.10
K_2O	1.37	4.10	5.90	4.10
H_2O	.76	.70	1.00	....
Total	99.91	101.20	100.10	101.10
Sp. Gr.	2.54	2.41	2.44	2.48

1. Soda rhyolite from Berkeley, California.
2. Liparite from Telkebanya, Hungary.
3. Rhyolite from Hot Springs Hills, Pahute Range, Utah.
4. Pantellerite, Kahania, Island of Pantelleria, Mediterranean.

Discussion of Analyses:

1. Rhyolites have the highest silica content of any volcanic rock and generally higher than granite.

2. They have low iron, magnesia and lime contents, due to the scarcity of dark minerals. The lime comes from the acid plagioclase.

3. Potash generally predominates over soda.

4. In pantellerite, soda predominates over potash, due to the presence of anorthoclase.

THE TRACHYTE FAMILY.

Mineralogical Composition.—Glassy orthoclase (sanadine). Biotite, hornblende, augite, diopside. Magnetite and titanite as common accessories. Volcanic equivalent of the syenites.

Texture.—Groundmass usually crystalline of sanadine, containing sanadine or orthoclase phenocrysts in which Carlsbad twinning is evident. Flow structure often conspicuous.

Relationship.—Trachytes pass into phonolites with increase in soda. They grade into syenites with the development of granitoid texture. They may be confused with andesites unless the striated feldspar of the latter is distinguishable.

Character.—The name trachyte is derived from a Greek work *trachus*, meaning "rough," because of the rough character of the first rocks of this type which were studied. They are not common, and are found in the following type localities: in the volcanic districts of Italy and the Auvergne, along the Rhine, in the Azores, in the Black Hills, in Custer County, Colorado, and in Montana.

Analyses of Trachytes (See under Phonolites).

Discussion of Analyses.—Compared with rhyolites, the trachytes show:

1. Lower silica, due to decrease in quartz.

2. Higher alumina, magnesia, lime and iron, due to increase in ferro-magnesian minerals.

3. Higher alkalies, potash usually predominating.

THE PHONOLITE FAMILY.

Mineralogical Composition.—Sanadine and nephelite or leucite. Aegirite. Occasionally members of the sodalite group. Garnet as an accessory.

Texture.—The groundmass is crystalline, sometimes porphyritic, rarely glassy.

Relationship.—Phonolite grades into trachyte with decrease in soda. The two types are closely associated.

Character.—Phonolite is a translation into Greek of a German word *Klingstein,* or "cluck stone," so named because certain phonolites with a pronouncêd horizontal jointing when hit give forth a metallic sound. The rock has a greasy appearance, due to the presence of nephelite. Nephelite if identified serves at once to distinguish phonolite from other volcanic rocks.

Leucite phonolites are rare. Leucite may and frequently does occur with nephelite in the typical phonolite. Concentrically arranged inclusions of magnetite specks occur in the leucite.

The pyroxenes are more common than in any other volcanic rock. They are usually ægirite-augite or ægirite in long-tufted, ragged, bright green prisms. The accessory minerals sodalite, brown garnet, and titanite are in themselves characteristic.

Phonolites are relatively not common. They occur in dikes, sheets and isolated buttes (Devil's Tower) in the Black Hills and in the Cripple Creek mining districts of Colorado, where they are associated with purple fluorite and calaverite in the ore bodies. The phonolite magmas being rich in alkalies may have had a solvent effect upon the gold, thus accounting for the present association.

In Germany, phonolites occur in great masses as volcanic necks or plugs in southern Baden, near the Swiss border. Many old castles have been erected on the summits.

Kilimanjaro, one of the volcanoes which has recently been active, is said to have given forth phonolite lavas.

Analyses of Trachytes and Phonolites:

	TRACHYTES.			PHONOLITES.		
	1	2	3	1	2	3
SiO_2 . . .	66.30	64.70	66.03	58.20	58.50	61.08
Al_2O_3 . .	17.80	16.50	18.49	21.60	19.70	18.71
Fe_2O_3 . . .	2.30	.70	2.18	21.80	31.40	1.91
FeO . . .	.40	2.70	.22	....	....	.63
MgO . . .	.30	1.70	.39	1.30	.30	.08
CaO . . .	2.10	3.20	.96	2.00	1.50	1.58
Na_2O . . .	5.60	2.70	5.23	6.00	10.00	8.68
K_2O . . .	3.50	5.50	5.86	6.60	4.70	4.63
H_2O . . .	.20	1.60	.85	2.10	1.00	2.21
Total . . .	98.50	99.30	100.20	97.56	99.10	99.51
Sp. Gr. . .	2.6	2.56	2.59	....	2.6	2.58

TRACHYTES.

1. Trachyte from Auvergne, France.
2. Biotite hypersthene trachyte from Tuscany, Italy.
3. Trachyte from Game Ridge, Custer County, Colorado.

PHONOLITES.

1. Phonolite from Schlossberg, Teplitz, Bohemia.
2. Phonolite from Miaune, France.
3. Phonolite from "Devil's Tower," Black Hills, Wyoming.

Discussion of Analyses.—Compared with trachytes, the phonolites show:

1. Lower silica due to the substitution of nephelite for sanadine.

2. Higher alumina and alkalies.

3. Lower iron, magnesia and lime, due to absence of dark minerals. In case ægirite is present, soda and iron are increased.

4. Traces of chlorine and sulphur are due to the presence of members of the sodalite group.

THE DACITE AND ANDESITE FAMILY.

Mineralogical Composition.—Acid plagioclase; biotite, hornblende, augite, diopside; magnetite, apatite, zircon as common accessories. Quartz is present in dacite and absent in andesite.

Texture.—Groundmass present as glass or as an intimate mixture of minute indistinguishable feldspars, which may be described as a "pepper and salt" texture. Plagioclase feldspar shows irregular outlines with zonal arrangement of inclusions frequent.

Character.—This group is the volcanic equivalent of the quartz diorite and diorite group. The dacites are not common. They were named from an old Roman province of Dacia, now a part of Hungary. Andesites derived their name from the abundance of lava of this type in the Andes Mountains.

Differentiation from other volcanic rocks may be based upon the peculiar "pepper and salt" texture. Dacite and rhyolite are confused unless the twinning of the plagioclase of the former is observed.

Among some of the active volcanoes which furnish andesite lavas are: Chimborazo, in the Andes; Aphroessa in the Santorin Archipelago, Aegean Sea, which was in eruption in 1863; Krakatoa, whose last eruption was in 1883, and the extinct volcanoes Mount Shasta, Mount Hood and Mount Rainier.

Classification of the Andesites:
1. Mica andesite.
2. Hornblende andesite.
3. Pyroxene andesite.
    a. Augite andesite.
    b. Hypersthene andesite.

Analyses of Dacite and Andesite (See next table).

Discussion of Analyses.—Compared with trachytes the analyses show:

1. Lower silica, due to the substitution of acid plagioclase for alkali feldspar.

2. Higher alumina, iron, magnesia and lime, due to the presence of the dark minerals.

3. Lower alkalies. Soda always predominates over potash. This is due to the presence of the acid plagioclase.

THE BASALT FAMILY, INCLUDING DIABASE.

Mineralogical Composition.—Basic plagioclase, augite; magnetite is a common accessory. Olivine is present in olivine basalt.

Texture.—These rocks possess a texture that is characteristic of rapid cooling. They have occasionally a glassy groundmass dotted with skeleton crystals, but more commonly they are porphyritic. The more crystalline portion of the rock consists of prominent augite and olivine crystals with good outline, and with plagioclase poorly developed in small crystals. The pale buff color of the augite phenocrysts is characteristic.

The texture of diabase is intermediate between that of gabbro and of basalt. It is essentially an intrusive basalt, entirely crystalline and granitoid. The plagioclase crystals are idiomorphic, occurring in long, lath-shaped crystals which lie in all positions. The interstices are filled with allotriomorphic crystals of augite and magnetite. This texture is called "ophitic." It is an important microscopic criterion for the identification of diabase.

Diabase is variously classified, sometimes as a plutonic rock and sometimes as a volcanic rock. Since it grades into porphyritic forms at the contacts and since it is really volcanic in its nature, occurring in sheets or dikes of limited thickness close to the surface, it is considered here with the basalts.

Basalts are volcanic equivalents of the gabbros. They are difficult to classify in the field, as they are all heavy, black gray or brown rocks for which a common and useful field term "dolerite" has been applied. The term diabase originated from the Greek verb *diabainein*, "to penetrate." Trap is a common field synonym applied to rocks of diabasic texture.

Classification.—A simple classification of the basalt family which meets all field requirements is:
1. Basalt.
2. Olivine-basalt.
3. Diabase.
4. Olivine-diabase.

Distribution.—Basalts are abundant particularly along the Atlantic seacoast where diabase has intruded Triassic

shales. It has formed prominent landmarks, such as the
Palisades of the Hudson, East and West Rock near New
Haven, and Deep River, North Carolina. Thousands of
feet of basalt of Pre-Cambrian age are found on Kewee-
naw Point. Native copper, secondarily precipitated in
the amygdaloidal cavities of these flows, is an important
ore. The Columbia Plateau and the Deccan Plateau furnish the two greatest examples of basaltic extrusion.

The lavas from many volcanoes are chiefly basaltic.
Among these are Kilauea and Mauna Loa in the Hawaiian
Isands, Mount Etna, and various volcanoes in Iceland.

Analyses of Dacite, Andesite, Basalt and Diabase:

	1	2	3	4	5	6
SiO_2 . . .	69.40	62 00	60.30	51.80	49.70	49.20
Al_2O_3 . . .	16.20	17.80	16.90	12.80	13.60	13.50
Fe_2O_3 . .	.90	....	5.90	3.60	7.80	5.50
FeO . . .	1.50	4.40	1.40	8.70	7.20	10.60
MgO . .	1.30	2.60	3.50	7.60	5.50	6.80
CaO . . .	3.20	5.40	5.60	10.70	12.40	11.50
Na_2O . .	4.10	4.30	3.80	2.10	1.60	1.80
K_2O . . .	3.00	1.50	2.40	.40	1.20	.10
H_2O . .	.40	1.70	.40	.60	.10	.30
Total . .	100.00	100.99	100.20	98.30	99.10	99.80

1. Dacite from Lassen's Peak, California.
2. Hypersthene andesite from Mount Shasta, California.
3. Augite andesite from Chimborazo, Mexico.
4. Diabase from New Haven, Connecticut.
5. Basalt lava from Thjorsa, Iceland.
6. Iron basalt from Nifak, Disco Island, Greenland.

Discussion of Basalt Analyses.—When compared with
andesites, the analyses show:

1. Lower silica.
2. Higher alumina.

3. Lower alkalies, all due to the increasing basicity of the feldspar.

4. Higher iron, magnesia and lime, due to the addition of dark minerals.

5. Soda predominates over potash.

TRACHYDOLERITES.

Mineralogical Composition.—Basic plagioclase and alkali feldspar; pyroxene; members of the sodalite group, olivine, and hornblende.

Texture.—Often porphyritic, with phenocrysts of basic plagioclase.

Relationship.—Trachydolerites are the volcanic equivalents of essexites. They are intermediate between alkali trachytes and phonolites on one side and tephrites on the other.

Analyses (See next table).—The analyses show low silica, high alumina, iron, magnesia, lime, and the alkalies.

TEPHRITES AND BASANITES.

Mineralogical Composition.—Basic plagioclase and either nephelite or leucite. Augite is common. Basanite contains olivine and tephrite does not.

Relationship.—These rocks are the volcanic equivalents of the theralites and the shonkinites.

Classification.—Basic plagioclase is common to all varieties.

1. Leucite tephrite, with addition of leucite and augite.

2. Leucite basanite, with addition of leucite, augite and olivine.

3. Nephelite tephrite, with addition of nephelite and augite.

4. Nephelite basanite, with addition of nephelite, augite and olivine.

The lavas of Mount Vesuvius are leucite tephrites, containing a little olivine, thus grading toward the leucite basanites.

Analyses (See next table).—The chemical character of this group is very similar to that of the last group. The alkalies are higher, due to the addition of leucite and nephelite.

LEUCITITES AND LEUCITE BASALTS.
NEPHELITITES AND NEPHELITE BASALTS.

Mineralogical Composition.—Leucite or nephelite with augite; with or without olivine. Nonfeldspathic. They grade from tephrites and basanites by the elimination of the basic feldspar.

Relationship.—These rocks are the volcanic equivalents of the ijolites and the missourites. The nephelite basalts accompany the phonolites in the Cripple Creek district of Colorado, and are sometimes completely impregnated with gold telluride.

Analyses (See next table).—The chemical character of this group corresponds closely to that of the tephrite and basanite group if the characteristics normally contributed by the basic plagioclase are deducted.

LIMBURGITE AND AUGITITE.

Mineralogical Composition.—Limburgite consists of pyroxene and olivine, and augitite consists of pyroxene only. They both contain apatite and magnetite as common accessories. They contain no feldspars nor feldspathoids.

Relationship.—These rocks are regarded as the most basic of the volcanic rocks and are the end products of the trachydolerite-tephrite-leucitite series. They may be regarded as the volcanic equivalents of peridotite and hornblendite.

Analyses.—Chemically they show a decrease in the alkalies, due to the absence of the feldspathoids. Limburgite shows an increase in magnesia, due to the presence of olivine.

	1	2	3	4	5	6	7
SiO_2 . .	51.76	48.09	47.40	45.90	46.90	42.78	43.35
Al_2O_3 . .	16.64	20.12	23.70	18.70	21.60	8.66	11.46
Fe_2O_3 .	14.06	6.72	6.80	....	8.10	....	11.98
FeO	....	4.32	3.50	10.70	....	17.96	2.26
MgO . .	3.21	4.19	1.90	5.70	2.50	10.06	11.69
CaO . .	8.15	9.37	6.50	10.60	8.00	12.29	7.76
Na_2O . .	4.98	2.62	6.40	1.70	8.90	2.31	2.88
K_2O .	1.31	5.69	3.30	6.80	2.60	.62	.99
H_2O . .	....	....	1.70	.60	2.10	3.96	2.41
Total .	100.11	101.12	101.20	100.60	100.10	98.64	95.80

1. Trachydolerite from Chajorra, Island of Teneriffe. Eruption of 1798.
2. Leucite tephrite from Atreo del Cavallo, Vesuvius. Eruption of May, 1855.
3. Nephelite tephrite from lava stream on San Antao, Cape Verde Islands.
4. Leucitite from Capo de Bova, Via Appia, Rome.
5. Nephelinite from San Antao, Cape Verde Islands.
6. Limburgite from Limburg, Kaiserstuhl Mountains, Baden, Germany.
7. Augitite from Hutberg, near Tetschen, Germany.

PYROCLASTIC ROCKS.

Pyroclastic rocks are those made up of fragmental volcanic deposits, usually more or less stratified by transportation through the air or under water.

Classification.—1. Volcanic agglomerate consists of the large sized volcanic products of a fragmental nature which have been deposited near the crater.

2. Volcanic breccia consists of angular fragmental products which have been firmly cemented together.

3. Tuff is a deposit of volcanic ash or dust consolidated by cementation. In the historical eruption of Vesuvius, in 79 A.D., Pompeii was buried in ash and Herculaneum was buried in tuff. Excavation has gone on rapidly in Pompeii, as the loose ash offers less resistance to removal than does the tuff which covered Herculaneum.

CHAPTER 11.

SEDIMENTARY AND METAMORPHIC ROCKS.

Sedimentary rocks are of secondary origin in that they are formed from previously existing rocks which may have been either igneous, sedimentary, or metamorphic. They may be mechanically or chemically deposited, either on land or under water. The agents of deposition are water, wind, and ice.

By the weight of overlying strata and through the agency of siliceous, calcareous or ferruginous cement, they become consolidated from a loose aggregate to a solid mass.

CLASSIFICATION.—1. Sediments of mechanical origin.

    A. Water deposits.

        a. Conglomerate.

        b. Breccia.

        c. Sandstone.

        d. Shales.

    B. Wind deposits.

        a. Loess.

        b. Sand dunes.

2. Sediments of chemical origin formed from solution.

    A. Concentration.

        a. Sulphates.

          Gypsum.

          Anhydrite.

- b. Chlorides.
  - Halite.
- c. Silica.
  - Flint, etc.
- d. Carbonates.
  - Limestone, etc.
- e. Oxides.
  - Iron Ores.

B. Organic,. through the agency of animals and plants.

- a. Carbonates.
  - Limestones of several kinds.
- b. Silica.
  - Diatomaceous earth, etc.
- c. Phosphate.
  - Phosphate rock.
- d. Carbon.
  - Coal, etc.

Sedimentary Rocks of Mechanical Origin.

Conglomerates.—Conglomerate is a rock consisting of cemented fragments of rounded, water-worn material of varying sizes. The pebbles are usually made of the more resistant varieties of minerals and rocks. Finer sediments fill the interstices. Conglomerates are aqueous in origin and show more or less stratification. The represent near-shore conditions of sedimentation.

Breccias.—A breccia is a rock composed of angular fragments cemented into a solid mass.

1. Talus breccia is derived by ordinary weathering and disintegration of a rock ridge. It accumulates at the base of slopes or cliffs.

2. Fault or friction breccia is derived from earth's movements, which crush the rock on two sides of a fault plane by friction or by intense pressure.

3. Volcanic breccia is formed by an accumulation of angular fragments ejected by volcanic action and later solidified.

Sandstone.—Sandstone is composed of sand grains which have been rounded by water action and separated by the classifying action of moving water to deposits of uniform texture. Quartz is the essential constituent, although impurities are always present, such as feldspar, mica, garnet, magnetite, etc.

CLASSIFICATION.—According to the character of the cement:
1. Siliceous sandstone.
2. Ferruginous sandstone.
3. Calcareous sandstone.
4. Argillaceous sandstone.

According to mineral content:
1. Arkose, containing much feldspar. ·
2. Graywacke, containing ferro-magnesian minerals.
3. Micaceous sandstone, etc.

Shale.—Shale is a rock consisting of the finer material, usually clayey, deposited beyond the zone of deposition of the sandstone. It contains compacted clays, muds and silts, which possess a finely stratified structure called bedding.

CLASSIFICATION.—According to composition.
1. Argillaceous shales.
2. Arenaceous shales.
3. Ferruginous shales.
4. Carbonaceous shales.

Shales form about 87 per cent of the sedimentary rocks, sandstones about 8 per cent, and limestones about 5 per cent.

Loess.—Loess is a fine, homogeneous, clay-like substance, largely siliceous, which lacks all semblance of stratification, and when eroded forms precipitous cliffs. It contains angular quartz, mica flakes, clayey material, with often high percentages of calcium carbonate.

Loess is believed to be a wind-blown deposit, probably assisted in certain localities by aqueous deposition. It is used in the West for brick manufacture.

Adobe clay is a form of loess abundant in the arid southwestern portion of the United States. It is used in the manufacture of sun-dried brick for adobe houses.

Sand Dunes.—Sand dunes are formed by wind-blown sand, which always exhibits a characteristic shape with its long, gentle slope on the windward side, up which the sand grains are blown, and with a steeper slope on the leeward side, which is the angle of repose for sand grains. Sand dunes show stratification and ripple marks.

The migration of sand dunes has been known to create havoc in certain parts of the coutry. They are "fixed" by transplanting with beach grass and sand hedges. One of the railroads temporarily checked the progress of some advancing sand dunes by spraying them with crude petroleum.

Sediments of Chemical Origin.

Gypsum and Anhydrite.—Gypsum and anhydrite, the hydrous and the anhydrous sulphates of lime, occur interbedded or in irregular masses, interstratified with clays, shales, sandstones, and limestones, or with halite.

They originate from concentration of oceanic waters by evaporation, and in inland lakes in which the evaporation equals or exceeds the amount of inflow.

Anhydrite changes to gypsum by normal hydration, due to exposure. A tunnel in Europe which was driven through anhydrite was thrown out of alignment by the volume increase produced by this change.

Halite.—Halite occurs in massive, granular form, interstratified with clays, marl and sandstone. It is especially associated with gypsum, anhydrite and dolomite. The deposits of Strassford, Germany, are 4,000 feet thick. It is here associated with the chlorides and sulphates of potassium and magnesium.

Flint.—Flint is a cryptocrystalline variety of silica occurring as a hard, grayish to blackish rock, its color being due to carbonaceous matter. It occurs as nodules or lenses in limestones. It is used for road material, and in tube and ball mills.

Sediments of Organic Origin.

Limestone.—Limestone is a widely distributed calcium carbonate rock containing impurities of magnesia, silica, clay, iron, and organic matter. It is quite soluble, and allows the formation of sink holes, caves, solution cavities, etc. Buildings erected of limestone thirty or forty years ago often show the effect of weathering by pitted and etched surfaces.

Limestone forms by chemical precipitation and through the agency of animals and plants.

CLASSIFICATION.—The classification of limestone is based upon composition, texture, and uses. It has a wide range of occurrence.

1. Calcic limestone is chiefly calcium carbonate.

2. Dolomite refers usually to any magnesian rich limestone.

3. Chalk is a soft, porous, fine-grained variety composed of minute shells of foraminifera.

4. Hydraulic limestone is a clayey limestone used in cement manufacture.

5. Lithographic limestone is a fine-grained, homogeneous variety used for lithographic work.

6. Oolitic limestone is composed of small spherical grains of calcium carbonate.

7. Travertine is a porous, cellular variety deposited by hot springs.

Iron Ores.—The oxides of iron, hematite and limonite, and the iron carbonate, siderite, may all have a sedimentary origin, either secondary or primary. They are all commercially valuable as a source of iron.

Phosphate Rock.—Phosphate rock consists chiefly of calcium phosphate. It has a value as a source of phosphoric acid in the manufacture of fertilizers. It is of organic origin, formed from animal remains. Large deposits are found in Florida, Tennessee, Idaho, Wyoming, and Utah.

Carbonaceous Rocks.—These rocks include all accumulations of vegetable matter that have undergone partial or complete decay under water.

The principal varieties which form transitional stages from the unaltered plant remains to graphite by steadily increasing metamorphism are peat, lignite or brown coal, bituminous or soft coal, and anthracite or hard coal.

Metamorphic Rocks.

Agents of Metamorphism.—The chief agents involved in the alteration of igneous and sedimentary rocks to their metamorphic equivalents are: (1) dynamic action due to earth movements producing shearing and folding of the rock formations, and (2) chemical action influenced by heat, liquids, and gases.

Composition of Metamorphic Rocks.—The chemical composition of metamorphic rocks is frequently similar to the composition of the rocks from which they are derived. In so far as this is true, chemical analysis is an important criterion for discriminating between metamorphosed sedimentary and metamorphosed igneous rocks.

Frequently there is addition or subtraction of constituents accompanying metamorphism which renders more difficult the interpretation of the origin of the metamorphic rock.

Mineral composition may or may not be the same in the metamorphic rock as it was in the original rock. Frequently metamorphism is accomplished by a granulation and rotation of the original particles. In the greater number of cases, there is a development of platy minerals which are best adapted to withstand conditions of higher pressures and temperatures. In these minerals the mutual parallelism of the greatest, mean and least dimensional axes causes a more or less perfect cleavage in one plane, which is called schistocity. The average ratio of the greatest to the mean dimensions of mica is 10 : 1, of hornblende 4 : 1, and of quartz and feldspar 1.5 : 1.

A metamorphic rock contains a higher percentage of the minerals mica and hornblende than the original rock. For example, shale may contain no mica. By metamorphism, mica schist is developed, containing over 50 per cent of mica. No change in chemical composition has taken place. Obviously the mica was developed by a recrystallization of the constituents originally contained in the rockmass.

Minerals which are characteristic of metamorphic rocks are staurolite, cyanite, sillimanite, zoisite, chlorite, talc, etc. Quartz, feldspar, mica, pyroxene, and amphibole are common to both igneous and metamorphic rocks.

Criteria for the Discrimination of Metamorphosed Igneous and Metamorphosed Sedimentary Rocks:

1. Mineralogical composition.

The minerals which are strongly indicative of a sedimentary origin of the metamorphic rocks in which they occur are staurolite, andalusite, sillimanite, cyanite. They all contain higher percentages of alumina than those found in igneous rocks, and as alumina is almost insoluble there is practically no possibility of an addition of alumina from other sources.

2. Original textures and structures.

If not too severely metamorphosed, sedimentary rocks may show remnants of bedding, fossils, cross-bedding, or other features. Igneous rocks may show amygdaloidal cavities, flow structure, etc.

3. Field relationships.

Areal distribution and association of the metamorphosed rock with surrounding rocks may give some clue. By tracing the metamorphosed rock laterally along the strike, one may come to a less metamorphosed portion of the rock which still shows original sedimentary textures or structures.

4. Chemical composition.

a. Dominance of magnesia over lime is indicative of sedimentary origin.

b. Dominance of potash over soda is suggestive of sedimentary origin.

c. The presence of several per cent of alumina over the 1 : 1 ratio necessary to satisfy the lime and alkalies is suggestive of sedimentary origin.

d. A high silica content is suggestive of sedimentary origin if supported by other criteria.

CLASSIFICATION.—The classification of metamorphic rocks is based upon composition, texture, and structure:

1. Gneisses.
2. Schists.
3. Quartzites.
4. Slates and phyllites.
5. Marbles.
6. Ophicalcite, serpentine, and soapstone.

TABLE OF SEDIMENTARY ROCKS AND THEIR
METAMORPHIC EQUIVALENTS.

Loose Sediments	Consolidated Rock	Metamorphic Rock
Gravel.	Conglomerate.	Gneiss, schist.
Sand.	Sandstone.	Quartzite.
Silt and clay.	Shale.	Slate, phyllite.
Lime deposits.	Limestone.	Marble.

TABLE OF IGNEOUS ROCKS AND THEIR
METAMORPHIC EQUIVALENTS.

Igneous Rocks.	Metamorphic Rocks.
Coarse-grained feldspathic rocks, as granite, syenite, etc.	Gneiss.
Fine-grained feldspathic rocks, as tuff, etc.	Gneiss, schist.
Basic igneous rocks, as diorite, basalt.	Schist.

Gneiss.—A gneiss is a banded metamorphic rock, either of igneous or sedimentary origin, in which the bands are mineralogically unlike, consisting chiefly of quartz and feldspar, with or without the parallel dimensional arrangement necessary for rock cleavage.

Gneisses are developed by a granulation, rotation and recrystallization of the original minerals rather than by the development of an entirely new set of the platy, cleavable minerals.

CLASSIFICATION.—According to the prevailing accessory mineral:

Biotite gneiss.
Muscovite gneiss.
Hornblende gneiss, etc.

According to origin:

Granite gneiss.
Gabbro gneiss.
Diorite gneiss, etc.

Schist.—A schist is a foliated, metamorphic rock whose individual folia are mineralogically alike, and whose principal minerals are the flat, platy minerals which are best adapted to withstand conditions of high pressure and high temperature. The parallel arrangement of the minerals develops the capacity to part along parallel planes, called schistosity.

CLASSIFICATION.—According to the prevailing schistose mineral:

Chlorite schist.
Mica schist.
Talc schist.
Actinolite schist, etc.

Quartzite.—Quartzite is developed from sandstone by a recrystallization of the original constituents into a hard, compact, crystalline mass having a splintery or conchoidal fracture.

A pure quartzite is rare, although the percentages of alumina, iron and the bases are often small. With an increase in impurities, the quartzite tends to take on a schistocity, due to the formation of the impure constituents by recrystallization into the flat, platy minerals. Quartzite schist is such a transition in which mica scales are found along the foliation planes. These planes probably represent original bedding planes in the sandstone.

Quartzite is used to advantage as a building stone, although its extreme hardness is found to be a handicap both in quarrying and in dressing.

Slate and Phyllite.—Slate is a dense, thinly cleavable, homogeneous rock, whose cleavage pieces are mineralogically unlike, and whose mineral grains are so small in size as not to be distinguished by the eye. This cleavage is not to be confused with original bedding planes. It is a secondary structure produced in the development of the secondary minerals.

Slates are composed of the finest particles of mineral matter which are carried in suspension and deposited considerable distances from shore. Volcanic ash and tuff more rarely give rise to slate deposits.

Phyllite is the next step in the metamorphism of a slate, intermediate between slate and mica or sericite schist. Quartz and mica are the essential minerals.

Marble.—Marble is the metamorphic equivalent of a limestone or a dolomite. It is completely recrystallized, and when pure shows the development of large rhombic calcite crystals or fine sparkling surfaces.

Few original limestones are pure. The metamorphism of an impure limestone containing silica, clayey material, iron oxides and carbonaceous matter is characterized not only by the recrystallization of calcium carbonate but by the development of various secondary silicates, particularly biotite, wollastonite, diopside, tremolite, actinolite, grossularite, and hornblende. At least seventy secondary minerals have been found to exist in metamorphosed limestones.

When pure, marble is massive, and shows no indication of a schistose structure.   All traces of fossils and original structures are obliterated.

Ophicalcite.—Ophicalcite is a variety of marble associated with streaks and spots of serpentine.   Verde antique is a name more popularly used.   It results from the metamorphism of an originally impure limestone to a calcite-silicate rock in which the silicates were later altered by hydration to serpentine.   Ophicalcite is valuable for decorative purposes, as it takes an easy polish. It occurs in quantities in Quebec, in the Green Mountains, and in the Adirondacks.

Serpentine.—Serpentine rock consists essentially of the mineral serpentine, a hydrous magnesium silicate, in association with olivine, pyroxene, hornblende, magnetite, chromite and the carbonates.   Garnets and micas are common accessories.

Serpentine is derived by metamorphism of igneous or other metamorphic rocks which are essentially composed of magnesium silicates, as olivine, pyroxene, or hornblende.   Such rocks are basic igneous rock and hornblende schist.

Serpentine occurs in the crystalline area of eastern United States, in eastern Canada, and in a few of the western coast States, but seldom in large masses.   It is used as an ornamental stone and as a source of asbestos.

Soapstone.—Soapstone is essentially the mineral talc. It becomes a talc schist by taking on a foliated structure. Impurities are mica, chlorite, tremolite, enstatite, magnetite, quartz, and pyrite.

Soapstone has a similar origin to serpentine as a secondary product from the magnesium silicates.   It is found in association with talcose and chloritic rocks in crystalline areas.

Soapstone is mined extensively in Virginia. The rock has many uses. It goes into the manufacture of tubs, switchboards, insulators, sinks, stoves, fire-brick and lubricants.

APPENDIX.

Suggestions for Geological Work.

The necessity for constant and careful observation cannot be too insistently urged in an examination of rocks or geological structures, whether it be a prelimiary reconnoissance or a final detailed survey of a property of limited extent. The geologist or mining engineer who is doing geological mapping should adopt the attitude that it may be impossible to return to the particular outcrop upon which he is working. Every feature of the rock which may be of possible value in the solution of the problem involved should be recorded on the spot. This outcrop may prove to be the keystone for the interpretation of the structure of the entire area. A close application of this rule will save much useful time and energy.

Observations for Geological Mapping.

The following outline (from Farrell) may be used as a guide to the geological features to be observed in an examination of a property:

A. RECONNOISSANCE OF THE AREA.

    1. Is the geology simple or complicated?

        a. Do the different formations cover a large or a small area?

        b. Is it easy to distinguish between them?

        c. Are the boundaries easy to find and follow?

    2. What are the probable rock types and their relations?

        a. Are rocks of igneous or sedimentary origin or both?

b. Are contacts conformable or unconformable?

c. In case of intrusive bodies, are there large dikes or small masses and few in number, or are they small and widely distributed through the intruded formations?

d. Are the rocks much altered?

e. Is metamorphism a prominent feature?

3. Collect specimens of the different formations, giving locations as closely as possible.

4. Note roads, trails, water, and possible camping places.

B. GEOLOGICAL MAPPING—GENERAL.

1. Locate boundaries between formations.

a. Simple boundaries.
Take dip and strike.
Does boundary indicate conformable or unconformable contact?
Are there evidences of faulting?

b. Obscure boundaries.
Look for fragmental traces of the formations.
Work up hill and locate the highest points at which fragments of the lower formation appear.
Note whether scarps or change of slope are connected with the boundary.

c. Complicated boundaries.
Intrusive boundaries.
Map carefully dikes and arms.
Note alteration and metamorphism in the neighborhood of the boundary.
Note variations in texture of the igneous rock in approaching the boundary.

Boundaries showing contact metamorphism.

Map the general relations of the metamorphic patches.

Note the metamorphic minerals and their succession.

Note presence and association of ore minerals.

2. Work within the boundaries of a formation. Traversing.

a. In areas of sedimentary rocks.

Strike and dip of beds.

Color, thickness and general character of beds.

Minerals composing the rocks; nature of the grains or fragments (angular or rounded) ; cementing material. In conglomerates look for recognizable fragments of earlier formations.

Presence of fossils.

Areas of alteration.

Areas of metamorphism.

Systems of folds—minor folding direction and pitches of axes of folds—relations of folding to faulting.

b. In areas of igneous rocks.

Rock texture and variations in texture.

Variations in composition.

Segregations.

Inclusions of other rocks.

Dip and strike of schistosity or gneissoid structure.

Flow structure.

c. In areas of metamorphic rocks.

Is rock of sedimentary origin?

Does it show traces of bedding?

Are gneissoid laminæ continuous, suggestive of sheared beds?

Are the grains rounded or angular in outline? What are their relative sizes?

Are minerals such as to suggest erosion or metamorphic processes?

Is the rock of igneous origin?

Are the minerals typical of igneous rocks?

Are the gneissoid laminæ noncontinuous, suggestive of sheared minerals?

Are the changes suggestive of dynamic action, chemical action or both?

Is the rock texture suggestive of folding and shearing?

Does it suggest an impregnation and metamorphism by replacement process, due to action of solutions?

Is the rock widely different in structure and composition from the original type?

C. GEOLOGICAL MAPPING—DETAILED WORK.

1. Faults.

   a. Strike and dip.

   b. Evidences of movement—slickensides, striæ (their direction and dip), gouge, drag, etc.

   c. Cross fracturing.

2. Veins and other ore bodies.

   a. Strike and dip.

   b. General character of mineralization.
      Strong or weak.
      Oxidized or unoxidized vein material.

   c. Minerals and groups of minerals.

   d. Relative age of minerals.

e. Represent exact outline of ore body as far as possible.

f. Note occurrence of branches or false walls.

g. Note character and extent of alteration of country rock.

h. Nature and extent of replacement of the wall rock by the ore.

Criteria of Relative Age.

1. Older rocks are more likely to have been deformed and metamorphosed, and therefore are harder to recognize.

2. Older formations are normally at the base of the series of sediments and flows.

3. Older formations are represented by fragments in later formations.

4. Younger formations fill erosion irregularities, fractures, fault planes and caves in older formations.

5. Younger formations cut the older ones as dikes, and include fragments of them.

Table for the Examination of Rocks in the Laboratory.

A. IGNEOUS ROCKS.

1. Texture, historical deductions, etc.

2. Mineralogical composition.

   a. Accessory minerals.

   b. Essential minerals.

3. Relative age of minerals.

   a. Minerals formed with crystal boundaries are older than the surrounding ones without crystal boundaries.

b. Included minerals are older than the ones which include them.

c. Minerals abutting without crystal boundaries are of the same age approximately.

d. Intergrown minerals are of the same age.

e. Minerals cutting others are younger than those they cut.

4. Alteration and metamorphism.

a. Degree of change, extent to which original minerals are changed.

b. Character of the change, secondary minerals due to alteration, metamorphic minerals.

B. SEDIMENTARY ROCKS.

1. Relative sizes and shapes of the component particles.

a. Unassorted material, large and small fragments together, imply that the source of the material is near at hand or that the transporting agent is very powerful.

b. Angular grains, fresh in appearance, indicate disintegration without decomposition, and little movement from the source.

c. Rounded grains, fresh, imply disintegration and transportation of the material.

d. Sorted material, where the grains are similar in mineralogical character, indicates that the deposits were made some distance from the source, or the original rock disintegrated and weathered also, differentiating the more resisting minerals.

   2. Look for fragments which give some clue as to the source from which the sedimentary material is derived.

   3. Determine the character and probable origin of secondary cementing material.

C. WITH METAMORPHIC ROCK TRY TO DETERMINE:

   1. The nature of the original constituents and the original rock. Look for traces of original minerals in form or cleavage.

   2. The nature of the alteration.

     a. Dynamic—folding, shearing, etc., distortion of crystals or fragments.

     b. Chemical change—older minerals partially dissolved by later ones.

INDEX.

Milton Keynes UK
Ingram Content Group UK Ltd.
UKHW021946030124
435382UK00004B/45

9 781015 559929